The Human Element in Advancing 21st-Century Space Exploration - Human Space Exploration Benefits to Humanity

Edited by Ilaria Cinelli, Aisha Said and Bettina Beard

Published in London, United Kingdom

The Human Element in Advancing 21st-Century Space Exploration - Human Space Exploration Benefits to Humanity
http://dx.doi.org/10.5772/intechopen.1006775
Edited by Ilaria Cinelli, Aisha Said and Bettina Beard

Contributors
Bashar Othman, Donna L. Roberts, Douglas Hamilton, Eleonore Poli, Francesca Grosso, Francesco Pagnini, Gabrielle Caswell, Ilaria Cinelli, Isaac Macharia, Kimia Seyedmadani, Leila Halawi, Lisa McNamee, Mark Miller, Peter H. U. Lee, Sam Holley, Sara Rothrock, Thobias Sarbunan, Ulpia-Elena Botezatu, Victor Vevera

First published in London, United Kingdom, 2026 by IntechOpen
IntechOpen is the global imprint of INTECHOPEN LIMITED, registered in England and Wales, registration number: 11086078, 167-169 Great Portland Street, London, W1W 5PF, United Kingdom

For EU product safety concerns: IN TECH d.o.o., Prolaz Marije Krucifikse Kozulić 3, 51000 Rijeka, Croatia, info@intechopen.com or visit our website at intechopen.com.

British Library Cataloguing-in-Publication Data
A catalogue record for this book is available from the British Library

The Human Element in Advancing 21st-Century Space Exploration - Human Space Exploration Benefits to Humanity
Edited by Ilaria Cinelli, Aisha Said and Bettina Beard
p. cm.
Print ISBN 978-0-85014-704-9
Online ISBN 978-0-85014-703-2
eBook (PDF) ISBN 978-0-85014-705-6

If disposing of this product, please recycle the paper responsibly.

Meet the editors

Ilaria Cinelli, Ph.D., FAsMA, FAsHFA, is a senior biomedical engineer specializing in space medicine, human factors, and space diplomacy. She holds degrees from the University of Pisa (Italy) and a doctorate in neural engineering from the University of Galway (Ireland), with additional training from the International Space University (The Netherlands) and Tufts University (USA). A Fellow of both the Aerospace Medical Association and the Aerospace Human Factors Association, she serves on the AsMA Council and the Mars Society Steering Committee, and formerly represented Europe on the IEEE Engineering in Medicine and Biology Society AdCom. Recognized among the 35 most influential women in space and designated a UN Space-4Women Mentor and Role Model, she was named one of 100 women changing the world by La Repubblica (2021) and honored as one of 50 tech voices in Europe by the European Patent Office (2023).

Aisha Said (she/her) is the Founder and CEO of Universal Inclusion, an organization dedicated to advancing diversity, equity, and accessibility in the space industry through podcasts, scholarships, and research. She also serves as Director of Programs at Space for Humanity (S4H), where she designs and leads the Citizen Astronaut Program, preparing participants for transformative spaceflight and guiding their impact projects on Earth. Aisha also chairs the S4H Inclusion Council, fostering best practices in equity and inclusion across the space sector. As a specialist in citizen and commercial spaceflight preparation, she draws on her degree in Inclusive Education to embed principles of universal design in her work, and on her degree in International Development Studies to bring a community-driven, globally minded perspective. Aisha is deeply committed to ensuring space exploration benefits all of humanity, both on Earth and beyond.

Tina Beard, Ph.D., began her career teaching undergraduate and graduate level courses and investigating human visual system plasticity. She then worked in the industry developing ophthalmological tests. In 1999, she shifted to government work serving in NASA management and as NASA Engineering and Safety Center (NESC) Deputy Human Factors Technical Fellow, leading efforts to: define human-automation function allocation strategies for the U. S. Air Transportation System, develop airline ramp controller training, design integrated workstations for tower controllers, and develop a methodology for aircraft inspector vision standards. Human spaceflight-related projects included an evaluation of NASA's test facility and payload procedures, ISS Mission Control Center resilience, astronaut workload, cognitive abilities after simulated space ascent, and normative data on NASA's Cognition Test Battery for spaceflight applications. In 2024, Dr. Beard retired from NASA and as chair of the Human Performance and Health Technical Committee of the International Association for the Advancement of Space Safety (IAASS).

Contents

Preface

Space exploration has long stood as one of humanity's most compelling frontiers. The upcoming retirement of the International Space Station before 2030 will mark the end of an extraordinary chapter in international cooperation, ushering in a new era defined by national ambitions, commercial expansion, and intensifying geopolitical competition.

Over recent decades, human spaceflight has fundamentally transformed what it means to be an astronaut. No longer viewed solely as superhuman figures, astronauts have become essential to the human, social, and strategic dimensions of space exploration. The growing presence of commercial astronauts continues to diversify the corps and reshape our understanding of who belongs in space.

As humanity prepares for permanent settlements on the Moon and Mars, and as technology approaches capabilities that may exceed human performance, we must ask: What does it truly mean to be an astronaut? What does it mean to be human in space?

The Human Element in Advancing 21st-Century Space Exploration – Human Space Exploration Benefits to Humanity presents diverse perspectives on the evolving role of people in shaping the next phase of exploration. Today's astronauts are not only explorers and technicians but also innovators, interpreters, and ambassadors who bridge scientific discovery, technological advancement, and societal engagement. They inspire future generations, foster global partnerships, and transform space research into tangible benefits for life on Earth.

This book examines how diversity, resilience, and innovation shape humanity's future in space, while considering the broader societal, cultural, and geopolitical implications of our expanding presence beyond Earth. It demonstrates that space exploration is a profoundly human adventure—one that reflects our shared aspirations, our capacity to collaborate across borders, and our determination to thrive in the harshest environments ever encountered.

We extend our sincere gratitude to the contributing authors for their thoughtful work in advancing dialogue across disciplines. Their contributions reinforce an enduring truth: the story of space will always remain, above all, a human story.

Ilaria Cinelli
Aerospace Medical Association
North Palm Beach, Florida, USA

Aisha Said
Space for Humanity
Boulder, Colorado, USA

Bettina Beard
International Association for the Advancement of Space Safety
The Netherlands

Chapter 1

Introductory Chapter: The Human Element in Advancing 21st-Century Space Exploration

Ilaria Cinelli

1. Introduction

Human beings have been explorers since the beginning of time. Our natural desire to explore, combined with our ability to adapt and our innovative spirit, has led us to discover new territories. Throughout history, we have evolved into more advanced beings who have built satellites, rockets, and other technological tools to extend our exploration capabilities.

As technological progress approaches the limits of human capability, a new chapter of space exploration is emerging, starting with lunar and Martian missions.

The upcoming massive space exploration push by multiple nations sparks a fundamental inquiry: Do we still need humans to explore outer space?

This book explores why the answer is yes. It looks at the unique value humans bring to space exploration, how our responsibilities are shifting in this new era, and what qualities—like perseverance, teamwork, and creativity—will remain essential for the journeys ahead.

2. Why humans matter in space exploration

Space exploration has always been about people. The ones who venture into the unknown, the experts behind the scenes who make it all work, and the rest of us who dream big and imagine what is possible. It is a journey that is deeply rooted in our nature, driven by a sense of courage, skill, and collective wonder.

The Cold War era introduced a new expression of this spirit through international superpower competition. National pride and prestige replaced exploration as the driving force behind discovery, which transformed the pursuit into an international competition of national pride.

April 12, 1961, was a day that changed everything. Yuri Gagarin, aboard Vostok 1, made history by becoming the first person to orbit the Earth. Gagarin's mission proved that humans can survive and operate in space, just like other explorers. A year later, President John F. Kennedy's speech at Rice University left a lasting impact. His words, "We choose to go to the Moon in this decade and do the other things not because they are easy, but because they are hard", sparked something. It turned the idea of reaching the Moon into a kind of goal for humanity, captivating people's imagination and setting the tone for years to come.

Shortly after the first human spaceflight, it was clear that the scientific nature of space exploration had transformed into a much more powerful force that extends beyond scientific investigation. The space programme evolved into a connecting element enabling international collaboration between space-based operations and Earth-based activities.

The International Space Station (ISS) represents the central achievement of such collaboration. Continuously inhabited since November 2000, the ISS functions as a living laboratory where scientific research meets technological innovation and diplomatic engagement. NASA astronauts, along with ESA personnel and Roscosmos cosmonauts, JAXA, CSA members, and representatives from private companies, research institutions, and emerging spacefaring nations, have conducted hundreds of missions, and more are yet to come [1]. The growing group of explorers becomes a symbol connecting government space initiatives and public support while representing the collaborative nature of contemporary space exploration. The astronauts who were once known as "superheroes" now represent the fundamental values of exploration, which include curiosity, ingenuity, diplomacy, and collaboration. The ongoing operation of the ISS is a living proof of how human teamwork—across borders, disciplines, and cultures—can achieve ambitious goals in science and discovery.

In more than two decades, the ISS has supported research investigations from over 100 nations, which have produced more than 2100 publications, including journal articles, conference papers, and technical reports that NASA tracks and updates through its "Station Results" platform [1]. The research conducted on the Station extends across multiple fields, including life sciences, materials engineering, Earth observation, and fundamental physics [1]. The experiments conducted by astronauts produce concrete and extensive benefits thanks to their enabling and interpretive capabilities [1]. For example, biomedical research conducted on the ISS has led to the development of Duchenne muscular dystrophy treatments, improvements in cancer immunotherapy, and advancements in mRNA delivery systems [2]. Human experimentation has enabled better early diagnosis of neurological diseases, including Alzheimer's disease, and it has developed wearable biosensors for managing chronic health conditions on Earth [2]. The ISS technology has enabled the development of filtration systems, which now serve more than 150 countries for clean water access [2]. The robotics technology developed for orbital operations now serves medical procedures in operating rooms, while ISS Earth observation systems enhance disaster response capabilities, environmental tracking, and global health monitoring [2].

Human activities in space directly produce these results. Astronauts perform three essential roles, which include serving as subjects and technicians while also functioning as innovators, interpreters, and mission enablers. The ability to adapt, think creatively, and make judgements becomes essential in spaces where unexpected challenges frequently appear or where technology cannot help.

2.1 A multipolar space era

The next phase of human spaceflight is now unfolding. Space agencies together with commercial enterprises and private investors direct their efforts toward establishing bases on the Moon and Mars and developing new platforms in low Earth orbit (LEO). The upcoming era will transform the current shared orbital laboratory operations into permanent independent settlements that exist beyond Earth's boundaries.

The transition brings forth intricate new operational dynamics. The presence of humans in space has evolved beyond scientific exploration because it now strongly connects with commercial activities and strategic and geopolitical factors.

2.2 Blurring boundaries: Civil, commercial, and military spaceflight

Since their beginning, defence operations and space exploration have been closely linked. The strategic interests of nations have elevated space into a vital domain for enhancing defence capabilities. National strategies will probably include human spaceflight as a method to achieve orbital security and defence-oriented objectives in the near future. The U.S. Space Force establishment, for example, shows how nations build specialised structures to unite space operations with defence strategies, although no crewed military missions exist at present [3]. NASA and the U.S. Space Force execute separate operations but maintain strategic alignment in their operational domain [4]. However, the role of astronaut was recently used to merge diplomatic research with security operations. NASA astronaut Mike Hopkins made history by becoming the first astronaut to join the U.S. Space Force when he transferred to the military branch during his ISS mission in 2020 [4]. Then, Colonel Nick Hague became the first Guardian to experience extended spaceflight during NASA's Crew-9 mission [5]. Taken together, this evidence may suggest that future human spaceflight could become a more visible instrument in bridging civil, diplomatic, and security interests in orbit.

The United States faces this trend as a single nation, but other spacefaring nations follow a similar path to establish space-based security frameworks. The primary focus of civilian agencies, like ESA, JAXA, and CSA, remains scientific discovery and exploration. However, multiple governments now establish programmes which combine scientific exploration with commercial and defence objectives for dual-use space-based asset utilisation [6, 7]. The development of these initiatives results from changing geopolitical conditions and commercial sector growth alongside the rising importance of space infrastructure. The dual-use orientation has also changed the way people view human spaceflight. The landscape is no longer limited to state-led missions, and now reflects a broader interplay between governmental, commercial, and societal actors, setting the stage for a new phase in which space activity expands beyond traditional boundaries.

The commercial spaceflight industry has introduced new exploration activities through suborbital flight operations. The brief nature of these missions enables less strict human participation standards, which creates opportunities for wider involvement. The space traveller population now encompasses private professionals, research experts, and ordinary citizens, in addition to government-selected astronauts [8–10]. This trend will likely continue to expand when preparing for future planetary missions. Analogue missions that duplicate planetary environments serve as the most successful method to prepare for extended space missions [11, 12]. Here, diversity in astronaut profiles helps to speed up research and exploration objectives.

The addition of various actors to space exploration brings fresh abilities together with new viewpoints and additional resources, which transform the conventional astronaut role. Astronauts perform two essential duties in modern times: exploring space and conducting experiments, and serving as ambassadors who link scientific research to governments, industry, and society through collaborative problem-solving. Human participation continues to define the path of space development through every mission that includes ISS expeditions and private space ventures.

2.3 Transitioning from the ISS to a decentralised landscape

The current period marks an essential transition for these developments. The ISS's retirement before 2030 will terminate the successful international cooperative platform that has operated for decades. The ISS operated as a political and diplomatic framework that supported joint missions, shared objectives, and mutual benefits between different cultures and nations for decades. The decommissioning of the ISS will establish a new space environment that national ambitions, commercial interests, and geopolitical tensions will define. China's Tiangong space station will stand as the sole government-operated orbital platform after the ISS retires, indicating a shift in institutional leadership. Commercial space stations developed through NASA and other agencies will start operations soon to reshape LEO access management and sustainability practices. The Outer Space Treaty upholds its fundamental principles yet encounters difficulties when applied to the modern transforming space environment. Human space exploration has relied on cooperative methods for 50 years, but the rising number of space actors with security, economic, and prestige interests threatens this approach.

2.4 The modern astronaut

Recent years have shown a substantial change in human involvement in space activities. The participation of multiple actors brings fresh perspectives and capabilities, transforming how we define an astronaut. Today's space crew encompasses commercial crew members, research scientists, analogue mission specialists, and maybe defence operators in the future. The various psychological, cultural, and operational needs of these profiles shape the development of mission plans, training programmes, and success assessment methods.

Involving humans in future exploration activities plays a vital role in developing planetary governance systems and space relations. The advancement of technology will not replace the need for astronauts to lead the human, social, and strategic development of space exploration.

The *Human Element in Advancing 21st-Century Space Exploration* book discusses that human beings continue to serve as the fundamental element in space exploration activities. The definition of the "right stuff" is changing, and the next generation of explorers will determine not only how humanity survives in space, but also how it thrives as a multi-planetary species. The story of space is, and will remain, a human story.

3. Why the future of space is human-centred

The following overview introduces the chapters of this book, each of which examines a different aspect of this evolving human dimension in space. Together, they highlight the challenges, opportunities, and transformations that lie ahead as we prepare for the next great chapter in humanity's journey.

3.1 Opening space to all: Diversity and opportunity in human exploration

Astronauts have inspired people on Earth to a degree that exceeds all expectations. The dream of becoming an astronaut has been a lifelong aspiration for many people throughout generations because this job remains one of the most sought-after positions. The exploration of space represents our most important values because it serves as an extraordinary human achievement, which pushes our creativity, determination, and teamwork skills to their limits under extreme conditions [13]. The same drive that motivated early humans to discover new continents and scale uncharted mountains manifests today through space exploration. Space missions enable us to advance technological frontiers while developing new capabilities that improve Earth-based life [13]. The experience of living and working in space requires us to face challenges, which lead to a better comprehension of human nature and our existence [13].

The astronaut's position as a science ambassador and inspirational figure has become more important in modern times [14]. Astronauts now use both conventional media and social media platforms to communicate science and educate people worldwide through their distinctive experiences. Through their physical presence, they connect people to Science, Technology, Engineering, and Mathematics (STEM) concepts, which motivates future innovators [14]. Their educational mission stands as a fundamental yet frequently ignored duty, which enables them to convert their experiences into a strong global outreach tool [14].

The qualifications required to become an astronaut need a total reevaluation for space exploration. During the Cold War era, the "right stuff" concept required both bravery and technical expertise from military test pilots. The evolving nature of space missions demands that astronauts fulfil different requirements because their responsibilities have grown more intricate. The modern "right stuff" consists of psychological and social resilience in addition to physical courage [15]. The future astronaut needs to possess exceptional adaptability, together with conflict resolution expertise and empathetic abilities. The astronaut needs to operate independently in space because Earth-based support becomes unavailable during extended missions [15]. The new requirements for space exploration require selection processes to expand beyond traditional astronaut profiles to include scientists, engineers, and commercial sector civilians [15].

The introduction of parastronauts challenges the traditional belief that space travel requires a specific physical appearance [16]. Space agencies and private companies expand their talent pool through mission modifications and technology adaptations for disabled individuals while demonstrating their dedication to diversity and inclusion [16]. The commercial space industry drives this change by offering both space tourism and private astronaut programmes. Virgin Galactic, SpaceX, and Blue Origin lead the way in transforming spaceflight from exclusive government control into a potential experience available to diverse individuals.

3.2 Human health: A frontier still to be explored

The basic purpose of governmental spaceflight demands continuous protection of astronauts' health and performance throughout their entire working period. Space agencies perform extensive medical and life science research to preserve crew health throughout their career.

The expanding number of new space actors creates changes to the existing space industry framework. The medical care and long-term health monitoring systems of commercial spaceflight operations and emerging ventures receive less attention than traditional space agencies. All space activities conducted by governments or private entities on lunar or Martian surfaces or other space locations should adhere to universal safety protocols.

However, significant obstacles are on the way when trying to adopt such principles outside government-led human activities. The achievement of astronaut safety and mission success requires substantial efforts in research across all human-centred disciplines to investigate risks and develop protective measures. The study of the human microbiome, for example, represents a priority in the field.

The microbiome functions as an invisible crew that controls digestion processes, immune system functions, and stress response mechanisms [17]. The microbial ecosystems in confined resource-limited environments such as spacecraft and planetary habitats experience changes that scientists have not yet fully understood. Small microbiome imbalances create problems that weaken immunity and boost disease susceptibility, and negatively impact mental health and cognitive abilities [17]. The development of probiotics and engineered food systems requires knowledge about how diet, stress, radiation, and closed-loop life-support systems affect the microbiome. The complex relationship between humans and microbiota and habitat ecosystems must be maintained for successful interplanetary missions. The prediction of microbiome changes enables the creation of sustainable life-support systems, medical protocols, and protective strategies for crew health during human space exploration [17].

3.3 Mission-critical minds

The preventive measures of space medicine have successfully reduced many risks and set priorities in developing solutions. The level of confidence in risk management has been reevaluated because future missions will build on new operations. The scientific community now focuses more on mental health risks because future extended space missions will introduce challenges that stem from the combination of prolonged isolation and confinement. Astronauts need training in stress management techniques to deal with the disruptions of their circadian rhythms, interpersonal conflicts, and the profound psychological impact of being distant from Earth. Small changes in crew cohesion or morale can trigger major operational problems, which negatively impact both mission performance and safety.

The exploration of space beyond LEO will result in extreme isolation and confinement, together with delayed communication, which will increase stress levels and make group interactions more complicated [18]. Human problems cannot be resolved through engineering solutions because they differ from technical malfunctions. The mission needs preventive measures to preserve mental health while developing teamwork capabilities and adaptability skills throughout its duration. The current astronaut training programme focuses on developing psychological resilience, together with conflict management abilities and high-pressure team collaboration skills [18]. Long-duration crews need team members who combine technical expertise with empathetic and flexible interpersonal support abilities because these qualities have become vital for mission success [19].

Earth-based analogue habitats serve as the foundation for deep-space mission preparation because both government space agencies and private companies use them. These facilities duplicate essential aspects of planetary life, which include isolation, confinement, and resource limitations that future space explorers will experience on their way to the Moon and Mars. The controlled and demanding environments allow scientists to study human responses to stress and limited external contact through the participation of astronauts and researchers. The acquired knowledge about crew cohesion and decision-making, as well as the effects of prolonged confinement, serves as essential information for developing spacecraft and habitats, and support systems that need to be resilient by design.

The analogue studies duplicate space conditions while exposing participants to the unanticipated difficulties of space missions. The research identifies potential risks in advance of orbital deployment while creating specific countermeasures, including medical protocols and optimised crew selection and training methods, and develops essential psychological and operational resilience for interplanetary exploration success.

The Outer Space and Earth Analogue Simulation (OASEAS) Habitat in Kenya represents a groundbreaking analogue astronaut research project that operates on the African continent [20]. Located in an equatorial semidesert region, OASEAS provides an environment that closely duplicates extraterrestrial conditions through its limited resources, harsh climate, and isolated conditions, which challenge participants. Researchers at this location can investigate the psychological and cognitive effects of isolation while also studying the operational aspects of supply management and crew health maintenance during extended periods [20].

3.4 Innovating for real: In space and on Earth

The understanding of human health preservation in LEO has advanced significantly during more than 20 years, but the same level of confidence remains elusive for Moon and Mars missions. The crew must operate autonomously from Earth-based support because evacuation is not possible, resources are limited, and Earth communication experiences delays. The debate about medical assistance and personnel involvement has become extremely intense. The "space surgeon" concept emerged because physicians-astronauts must provide all medical requirements from routine care to urgent surgery under the unique physiological and psychological conditions of deep-space missions [21]. Autonomous diagnostics work together with robotics and artificial intelligence to support crew health, but these technologies lack the human capabilities of judgement, adaptability, and empathy.

The main obstacle to successful human spaceflight innovation stems from the complex relationship between psychological and operational resilience and other factors [22]. The challenge has led to an urgent need for better space technology to support human life-support systems. The development of autonomous medical solutions and deep-space mission life-support systems requires astronaut health and operational resilience, which drives innovations with broader applications [22]. The requirements of extended missions develop technologies, protocols, and professional skills that improve healthcare delivery, educational systems, and human resilience capabilities on Earth. This development requires the development of new professionals who can merge STEM expertise with human-centric skills.

The technological and operational requirements demonstrate why it is essential to train future professionals who link STEM knowledge with human-oriented abilities, including communication, collaboration, and critical thinking [23]. The implementation of space technology innovations within educational systems through project-based learning and augmented and virtual reality approaches helps students build spatial reasoning abilities and design thinking competencies through real-world practical experiences. This method of workforce development will produce the following generation of technological and scientific advancements [23]. The protection of astronaut health, cohesion, and adaptability serves as a mission requirement that drives innovation, which benefits both space exploration and Earth-based humanity.

4. Conclusion

Space exploration activities depend on the human element, which advances dynamically while maintaining its fundamental importance for all space exploration activities. Space exploration has transformed from its origins as a Cold War government-trained test pilot competition into a worldwide system that combines international partnerships with commercial space ventures and astronauts from diverse backgrounds. The technological progress we develop today transforms both our space exploration objectives and the core principles of our space missions.

The following chapters of this book will explore the essential topics that were introduced in this chapter, including deep-space mission psychological and medical resilience, and the significant societal and economic effects of space presence. The human element's challenges and innovations, together with its enduring value, will provide us with a better understanding of space exploration's future direction. Our cosmic exploration reflects our human potential and dreams as we establish new definitions for a resilient multi-planetary species.

Acknowledgements

The author extends deep appreciation to the Aerospace Medical Association (AsMA) because this organisation inspired the book's development.

The nonprofit U.S.-based organisation AsMA creates a distinctive platform for specialists who work in industry, government, and academia to exchange knowledge and develop human-centred aerospace disciplines. The organisation has made vital contributions to aviation and space medical standards as well as flight safety and adaptation to extreme environments. The association's worldwide understanding, together with its specialised knowledge, significantly influenced and motivated this research.

Conflict of interest

The authors declare no conflict of interest.

Author details

Ilaria Cinelli
Aerospace Medical Association, Alexandria, VA, USA

*Address all correspondence to: i_cinelli@yahoo.it

References

[1] NASA. International Space Station Benefits for Humanity 3rd Edition [Online]. 2022. Available from: https://www.nasa.gov/humans-in-space/entering-the-decade-of-results-international-space-station-benefits-for-humanity-publication-released/ [Accessed: 1 August 2025]

[2] NASA. International Space Station Benefits for Humanity 2022 [Online]. 2022. Available from: https://www.nasa.gov/wp-content/uploads/2022/09/iss_benefits_for_humanity_2022_book.pdf?emrc=170cd1 [Accessed: 1 August 2025]

[3] Space Force. Space Systems Command—Command Plan [Online]. Available from: https://www.ssc.spaceforce.mil/About-Us/SSC-Command-Plan [Accessed: 1 August 2025]

[4] Almeida A. NASA Astronaut Mike Hopkins Transfers to US Space Force While Aboard International Space Station [Online]. Available from: https://www.spaceforce.mil/News/Article/2452899/nasa-astronaut-mike-hopkins-transfers-to-us-space-force-while-aboard-internatio/

[5] NASA. Nick Hague [Online]. Available from: https://www.nasa.gov/people/nick-hague/ [Accessed: 5 August 2025]

[6] European Union. European Defence-Readiness 2030 [Online]. 2025. Available from: https://eda.europa.eu/news-and-events/news/2025/03/19/joint-white-paper-for-european-defence-readiness-2030 [Accessed: 5 August 2025]

[7] General Secretariat of the Council of the European Union. Space: The New Battleground of Modern Warfare [Online]. 2025. Available from: https://espas.secure.europarl.europa.eu/orbis/document/space-new-battleground-modern-warfare [Accessed: 5 August 2025]

[8] Griko YV, Loftus DJ, Stolc V, Peletskaya E. Private spaceflight: A new landscape for dealing with medical risk. Life Sciences in Space Research (Elsevier Ltd). 2022;**3**:41-47. DOI: 10.1016/j.lssr.2022.03.001

[9] di Giulio I, Anderton R, Caplin N, Hodkinson P, Miller-Smith MJ, Narici Mv, et al. Preliminary considerations for accessible space missions for all. In npj Microgravity. Nature Research. 2025;**11**:1-5. DOI: 10.1038/s41526-025-00494-9

[10] Overbey EG, Ryon K, Kim JK, Tierney BT, Klotz R, Ortiz V, et al. Collection of biospecimens from the inspiration4 mission establishes the standards for the space omics and medical atlas (SOMA). Nature Communications; 2024;**15**(1):41-4. DOI: 10.1038/s41467-024-48806-z

[11] Van Ombergen A, Rossiter A, Ngo-Anh TJ. 'White Mars'—Nearly two decades of biomedical research at the Antarctic Concordia station. Experimental Physiology. 2021;**106**(1):6-17. DOI: 10.1113/EP088352

[12] NASA. Analog Missions [Online]. Available from: https://www.nasa.gov/analog-missions/ [Accessed: 2 August 2024]

[13] Halawi L, Miller M, Holley S. The human element in space exploration: Embodying our highest values. In: The Human Element in Advancing 21st-Century Space Exploration. IntechOpen; 2025

[14] McNamee L. Astronauts as STEM and terrestrial science ambassadors. In: The Human Element in Advancing 21st-Century Space Exploration. IntechOpen; 2025

[15] Roberts DL. The new 'right stuff'—What it takes to be an astronaut of the future. In: The Human Element in Advancing 21st-Century Space Exploration. IntechOpen; 2025

[16] Poli E. Diversity and inclusion in space-the case for parastronauts. In: The Human Element in Advancing 21st-Century Space Exploration. IntechOpen; 2025

[17] Hamilton D, Caswell G. Travelling companions: The human microbiome document details. In: The Human Element in Advancing 21st-Century Space Exploration. IntechOpen; 2025

[18] Pagnini F, Grosso F. The psychological needs of future astronauts Document Details. In: The Human Element in Advancing 21st-Century Space Exploration. InTechOpen; 2025

[19] Botezatu U, Vevera A-V. Resilient by design: Mitigating human-centered risks in space missions and critical infrastructure. In: The Human Element in Advancing 21st-Century Space Exploration. IntechOpen; 2025

[20] Macharia I. Pioneering analog astronaut research in Africa: The OASEAS Habitat in Kenya. In: The Human Element in Advancing 21st-Century Space Exploration. IntechOpen; 2025

[21] Othman B, Rothrock S, Lee P. Beyond the algorithm: The irreplaceable role of human doctors in space. In: The Human Element in Advancing 21st-Century Space Exploration. IntechOpen; 2025

[22] Seyedmadani K. Technology and innovation driven by space. In: The Human Element in Advancing 21st-Century Space Exploration. IntechOpen; 2025

[23] Sarbunan T. Integrating space technology innovations into holistic educational frameworks: Bridging STEM and human-centric skills. In: The Human Element in Advancing 21st-Century Space Exploration. IntechOpen; 2025

Chapter 2

Perspective Chapter: The Human Element in Space Exploration – Embodying Our Highest Values

Leila Halawi, Mark Miller and Sam Holley

Abstract

SpaceX mission and the Polaris Program extended astronaut competencies and skills, suggesting mixing traditional practices with pioneering methods. The current astronaut symbolizes a progression, integrating old skills with evolving competencies. While maintaining core technical and scientific aptitudes, today's astronauts or space explorers must demonstrate mastery in autonomous operations, cross-cultural collaboration, and navigation between governmental and commercial space protocols. Current training practices echo this change, emphasizing resilience for prolonged isolation, self-governing management proficiencies, and competence throughout distinct spacecraft systems. This chapter explores the fundamental human qualities that drive successful space exploration: resilience, curiosity, collaboration, and hope. We highlight how these vital traits facilitate mission success and technological innovation by analyzing astronaut training, mission experiences, and personal transformation. Understanding these human factors becomes vital for future mission design and crew selection as humanity stands at the threshold of permanent lunar presence and Mars exploration.

Keywords: space exploration, resilience, curiosity, collaboration, hope, safety, standards

1. Introduction

In the modern day of space exploration, high-tech and individual competency have redefined the borders of outer space. As space missions progress from short orbital flights to prolonged lunar stays and possible Mars expeditions, the role of astronauts has changed significantly.

Programs like NASA's Artemis program [1] and Mars expeditions mandate unique human adaptability and skills. These aspiring Moon-to-Mars missions [2] require a unique generation of space explorers who should excel in complex and self-directed tasks. This evolution was unexpectedly exemplified in 2024 when what commenced as a brief June expedition to the International Space Station transformed into an extended orbital residence for two NASA astronauts piloting the Boeing Starliner during a flight evaluation [3]. This unanticipated long-duration mission is a crucial real-world case study for NASA programs requiring exceptional human adaptability.

The unplanned extension provides valuable insights for future missions that demand unprecedented crew expertise and resilience.

With the SpaceX mission and the Polaris Program, astronaut skills mixed conventional practices with new methods. Present astronauts represent an evolution, incorporating technical and scientific skills with evolving ones supporting their ability to master autonomous operations, cross-cultural collaboration, and navigation between governmental and commercial space protocols. Space exploration illustrates people's greatest aspirations, symbolizing basic principles of resilience, curiosity, collaboration, hope, and an ethical obligation. These values drive our initiatives beyond the confines of Earth and are honed through the distinct challenges faced within the cosmic span.

This chapter presents the evolution of astronauts' skills, emphasizing the significance of resilience and curiosity's key role as a catalyst for scientific progression. We examine collaboration's value beyond boundaries and foster hope that inspires future astronauts. Finally, we ponder our commitment to safeguarding safety protocols and standards.

2. The evolution of astronaut skills: A chronological perspective

A noteworthy evolution accompanied the progression of human spaceflight in the indispensable skill sets of astronauts who voyage past Earth. It reports a fascinating story about people's shifting connection to space and advocates a trajectory for selecting and training astronauts.

From the 1960s to the 1980s, space agencies predominantly favored military test pilots. The first humans in space aboard Vostok 1 in 1961, Yuri Gagarin and Alan Shepard, the first American to reach this momentous event in the same year, demonstrated peculiar skills in flight operations [4]. John Glenn, who successfully orbited the Earth in Friendship 7 in 1962, exemplified the critical importance of test pilot accuracy during system failures [5].

As space programs evolved from the 1980s to the 2000s, a necessity for a more diverse range of expertise emerged. This era signaled the rise of scientist-and engineer-astronauts. The medical doctor and engineer, Story Musgrave, who participated in shuttle missions, represented this unique model [6]. Also, in 1983, Sally Ride, the first American woman in space with her physics acumen, fulfilled complex mission objectives [7].

From the 2000s forward, astronauts' credentials extended to the Versatile Explorer model. Space explorers exhibit psychological endurance and physical adaptability to endure prolonged isolation in challenging settings. Jessica Meir's involvement in the 2019 all-female spacewalk emphasized the significance of thorough training [8].

Christina Koch, with the record for the most extended duration of female spaceflight at 328 days, exemplifies the resilience requisite for deep space missions [9]. The latest cohorts of astronauts exemplify an increasing emphasis on diversity. NASA's most recent class comprises Nichole Ayers, Marcos Berríos, Chris Birch, Deniz Bunham, Luke Delaney, Andre Douglas, Jack Hathaway, Anil Menon, Chris Williams, and Jessica Wittner, who undertook training in conjunction with UAE astronauts Mohammad AlMulla and Nora AlMatrooshi [10].

Simultaneously, commercial astronauts have notably emerged. The SpaceX Crew-9 mission includes NASA astronaut Nick Hague and Roscosmos cosmonaut Aleksandr Gorbunov [11]. The Polaris Dawn mission, spearheaded by entrepreneur Jared Isaacman, aspires to execute the first commercial spacewalk with a civilian crew [12].

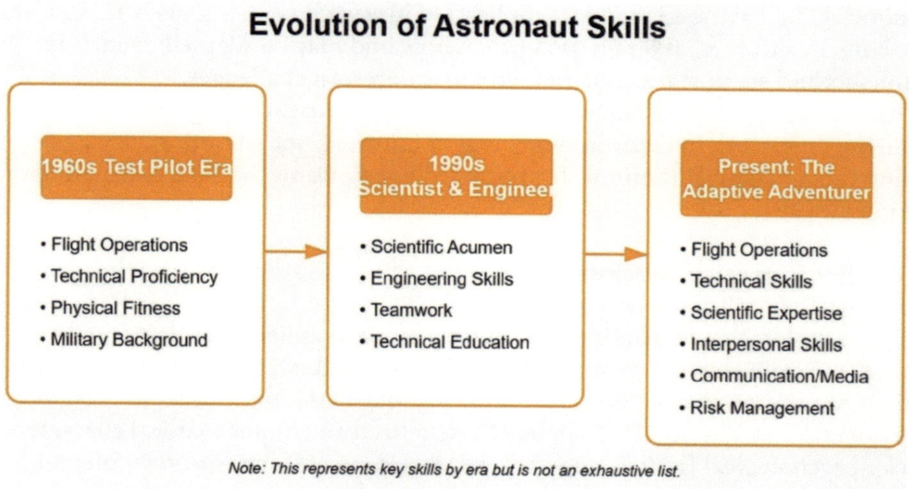

Figure 1.
Astronauts' skills evolution.

Despite originating in distinct operational frameworks, government and commercial astronauts show increasing convergence: government astronauts operate in conservative, oversight-intensive environments with comprehensive training and standardized procedures, while commercial astronauts function in efficiency-driven settings with specialized expertise, flexible procedures, and market-aligned risk tolerance. Contemporary space operations demonstrate significant paradigmatic integration. NASA now actively engages civilian astronauts through the Commercial Crew program [CCP], incorporating private-sector methodologies into government missions. This bidirectional exchange enables governmental adoption of commercial agility while private enterprises integrate government-derived safety protocols.

This sequential progression from test pilots to scientist-engineers to versatile explorers reflects the broader path of space exploration, where we can see some overlap within operational paradigms emerging between governmental and commercial space sectors. In **Figure 1**, we highlight the evolution of the astronauts' skills, showcasing how the field evolved to include a greater breadth of human abilities beyond technical proficiency, bearing in mind the inherent overlap Neither this figure nor the accompanying list should be considered exhaustive, as they represent selected examples within the complex continuum of astronautical competency development.

3. Resilience in adapting to extreme environments

Resilience, a versatile belief with longstanding origins, is a vital skill for astronauts in our present. Resilience is adapting, recovering, and thriving despite hardship, uncertainty, and high-stakes challenges. Although conventional definitions of resilience generally consider reactive abilities [13], recent definitions highlight the significance of endurance, mending, and redirection in disasters [14].

In space exploration, resilience is maintaining psychological and physical stability while navigating space missions' isolation, confinement, and inherent risks [15, 16]. Frameworks like the Resilience Assessment Grid (RAG) that assess the 'potential to anticipate, the potential to respond, the potential to learn, and the potential to monitor" [17] "have seen practical application in several contexts, such as in railways, off-shore, health care, and radiation protection, and may be adapted to enhance astronauts and support team adaptability to unforeseen challenges.

Astronauts apply resilience by managing stress, maintaining focus, and fostering teamwork in environments where failure is not an option, such as during technical malfunctions or prolonged missions on the International Space Station [18, 19].

3.1 Resilience in space missions

Space exploration demands extraordinary human resilience, as demonstrated through missions that progressively expanded our understanding of human adaptation. Resilience is not static but evolves throughout a mission. Our focus is on four core dimensions: Physical [human body's structural and biomechanical characteristics], psychological [human cognitive processes, mental states, and emotional responses], Operational [how humans perform tasks and interact with systems within a specific context or environment to achieve goals.] and Physiological [internal functioning of the human body and its responses to work demands and environmental conditions.], as illustrated in **Figure 2**, stems from their fundamental integration with established human factors principles. These dimensions represent the essential interfaces between humans and their environment, capturing the external challenges astronauts face and their internal responses to those challenges.

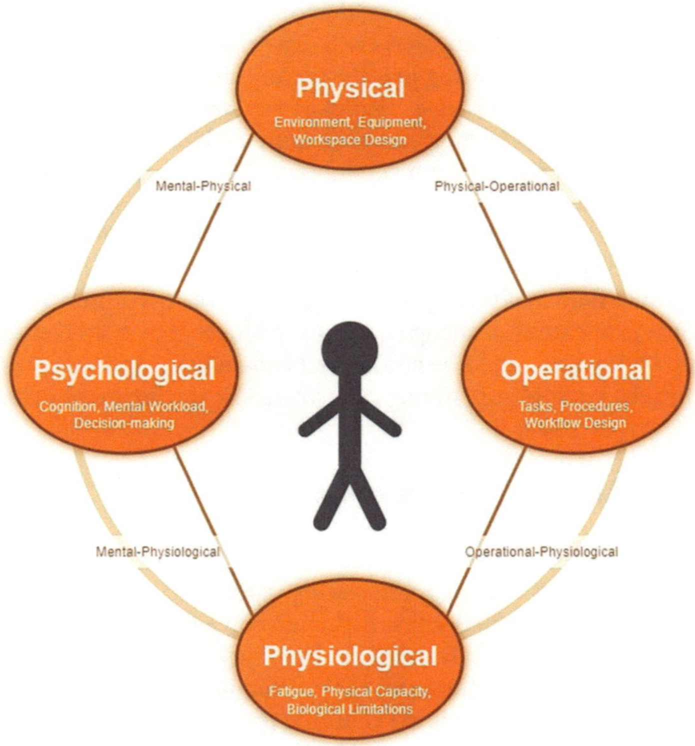

Figure 2.
Dimensions of astronauts' resilience.

The development of this resilience begins long before launch. During training, astronauts endure simulated crisis scenarios, prolonged isolation exercises, and physically demanding preparations.

Soviet cosmonaut Dr. Valeri Polyakov's 437-day Mir mission (1994–1995) established foundational evidence for long-duration spaceflight capability [20]. Extended spaceflight's psychological and physiological stresses demand resilience from astronauts [21]. Research emphasizes the importance of preparing astronauts for isolation and the stresses of deep space travel [22]. The Mir space station (1986–2001) became a crucial testing ground for human resilience, hosting 125 astronauts from 12 nations and supporting 28 long-duration expeditions. Despite technical challenges, including fires and collisions, Mir's 15-year operation tripled its projected lifespan, advanced space habitation knowledge through 23,000 experiments, and strengthened U.S.-Russian cooperation.

The 2013 spacewalk emergency involving Commander Chris Hadfield is a compelling empirical case study illustrating the practical application of "resistance, recovery, and reorientation" [14]. Hadfield's handling of water accumulation in his helmet during an ISS extravehicular activity (EVA_, which threatened his health and life support, exemplified operational resilience through resistance, recovery, and reorientation [23]. The psychological resilience necessitated by prolonged spaceflight represents another complex challenge, as exemplified by Scott Kelly's landmark 340-day mission aboard the International Space Station, which considerably enhanced our comprehension of extended space habitation through his application of adaptive strategies such as rigorous daily routines, consistent exercise regimens, and meaningful engagement activities, including the cultivation of the first space-grown flowers thereby establishing empirical foundations for psychological support protocols that will inform future Mars expeditions [24].

3.2 Mission control onboard AI CRM model for resilient, safe space

Resilience for space operations beyond the moon and, in particular, long-term space flight to places such as Mars will test astronauts as these lengthy flights will require some teamwork between humans and their spacecraft machines. At this juncture, it will have to be assumed that any such lengthy human-crewed flight in a spacecraft to the red planet, Mars, or others will require a great deal of computer automation. However, artificial intelligence [AI] also presents a great deal of computer information, and then teams up with the crew to make decisions. In the past, NASA's Mission Control was in constant contact with the astronauts, which accomplished this decision-making process. However, the distances that future astronauts will travel will require constant updates from an AI Mission Control on board the spacecraft. In light of this, NASA's role would shift to Mission Support, and Mission Control functions would transition to AI-human teaming on spacecraft. This type of AI Mission Control will require the astronauts to constantly interface with it because the computing power behind it will enable it to constantly gather detailed information about the navigation of the voyage, everything within the spacecraft operating, and everything ongoing with its human occupants and their resilience welfare (physically, physiologically, psychologically). With this human-AI interaction becoming a team, it is only reasonable to assume there will be a need for a standardized teamwork method for the human-machine-AI interaction based on the need to retain astronaut resilience on such lengthy space flights. To accomplish this, it is recommended that a new crew resource management [CRM] model for space flight be adopted that can

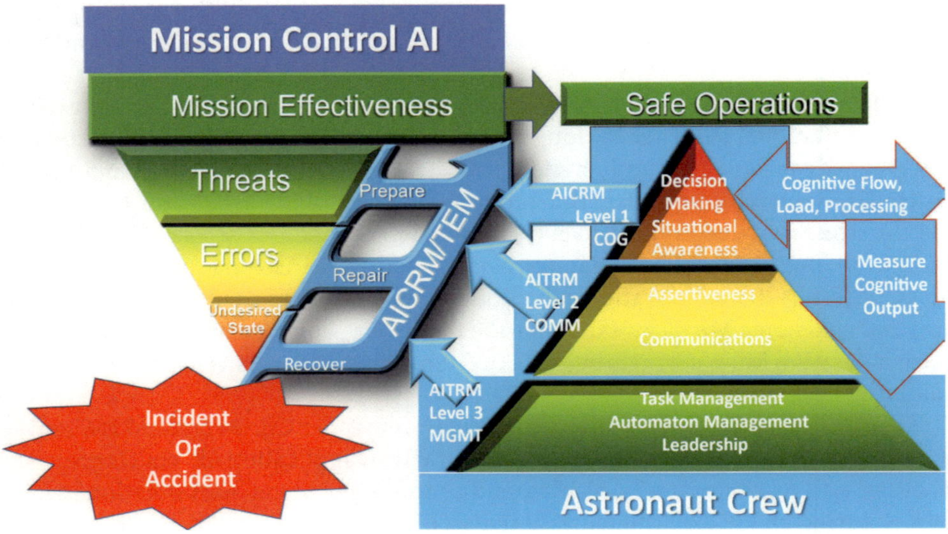

Figure 3.
Mission control onboard AI CRM model for resilient, safe space flight operations, adapted from CRM/TEM skills model [25] with enhanced Digital Flight Deck CRM/TEM skill level hierarchy and Digital Flight Deck Management for AI-Human Team Decision Making.

integrate basic tenets of current CRM into a hierarchy model of cognition that attains the highest situational awareness levels to make the best space flight decision, teaming with the onboard Mission Control AI system. To ensure the highest levels of safety are accounted for, a risk management model will be built into the AI system in the form of a threat error management model to update the human crew constantly. Such a model could be perceived as looking like the one in **Figure 3**, adopted initially from US Naval Aviation [26], but then having the tenets adjusted for a cognitive hierarchy for better situational awareness and decision-making while interacting with the AI and the Threat and Error management [TEM] status [25].

Domain-specific parameters currently limit the evolution of AI cognitive functions. The more complex functions emulating human cognition are not developed to withstand the rigors of extended space flight and solar system exploration. On long-duration space missions, machine learning opportunities and robotic teamwork processes may likely achieve a Mission Control System enhanced by experience gained while underway. The current consensus is that any such system would require a human in the decision loop in some capacity [27]. Connecting with TEM, AI, and established behavioral markers is paramount in this light. Exigencies of space travel and known unknowns or unknowns will be unique to that mission, at least in the early phases. The effects on cognitive loading and processing may be profound, especially when unanticipated situations arise. This will require a robust learning environment and effective machine-human teaming [28]. Integrating the AI CRM model would materially aid the human crew and the mission objectives.

4. Curiosity as a catalyst for scientific discovery

Despite geopolitical and economic factors, curiosity drives space exploration. Garcia Llama alludes to the concept when studying how humans "cross borders and extend their presence to other areas" [29]. The Artemis Program remains a current demonstration of this curiosity. Its aspiring lunar base initiatives surpass simple flag-planting symbolism, visualizing sustained human presence that ultimately incorporates astronauts and civilian sightseers [30]. NASA's work with Sierra Space further illustrates how practical innovation emerges from fundamental questions about our capabilities beyond Earth. Their Dream Chaser spacecraft represents more than just transportation; as Miguel Pereiro points out, it embodies the drive to ensure "more American spacecraft (and not just SpaceX) carrying cargo to the ISS." Behind the technical specifications lies a more profound curiosity about expanding human operational reach in orbit.

When we examine the trajectories of individual astronauts, the varied expressions of scientific curiosity become even more apparent. Dr. Kalpana Chawla's intellectual journey began with childhood wonder about flight, a common enough starting point. However, her genius was transforming that initial fascination into methodical engineering contributions at NASA. The Robotic Situational Awareness Display she developed reflected her technical expertise and curiosity about how humans and machines might better interact in space. Her historic 1997 Columbia mission was not merely about being the first Indian woman in space; it represented the culmination of questions she had pursued throughout her career. By her final 2003 mission, Chawla oversaw experiments ranging from crystal formation to cancer research, diverse inquiries tragically interrupted during Columbia's re-entry [31].

Don Pettit approaches space research with a different temperament. Where some follow rigid experimental protocols, Pettit embraces spontaneous investigation through his "science of opportunity" approach. This is not haphazard tinkering; it is curiosity responding to immediate circumstances. When Pettit observed unusual freezing patterns in water, designed the "Space Cup" using fluid dynamics principles, or demonstrated electrostatic orbits with knitting needles and water droplets, he engaged in scientific jazz—improvising within established principles to discover new patterns. His work reminds us that curiosity often thrives in unplanned moments [32].

Dr. Story Musgrave brings yet another dimension to space exploration. Through six shuttle missions, Musgrave cultivated an unusual perspective combining technical precision with artistic sensitivity. His documentation of Hubble Space Telescope repairs offered insights into the complex relationship between humans and technology in the cosmic environment [33, 34]. By bridging space work's mechanical and experiential aspects, Musgrave showed how curiosity encompasses objective observation and subjective experience [6].

Looking forward, NASA recognizes the evolving challenges of space exploration as systems incorporate increasingly sophisticated technologies. Their acknowledgment that "crew vehicle interactions will become more complex and interactive" points to how curiosity must adapt to environments with varying levels of automation. The questions we ask about space change as our technological capacities evolve, a recursive relationship between curiosity and capability that defines the ongoing human venture beyond Earth.

5. Collaboration across national and organizational boundaries

Space exploration entails constant cooperation beyond national borders [35], while historical junctures have actively redefined international cooperation through collaborative manifestations.

The Apollo 13 crisis in 1970 established a precedent for international cooperation during space-related emergencies. An oxygen tank failure during a space mission 200,000 miles from Earth triggered a transnational engineering collaboration for crew rescue. This crisis, which involved Soviet aid during the Cold War, embodied space exploration, promoting cooperation past geopolitical pressures [36].

Stranded cosmonaut Sergei Krikalev built global connections through radio while enduring isolation on Mir for 10 months [37].

SpaceX's collaborative public-private partnerships drove reusable rocket technology, transforming space economics and culminating in the historic May 2020 Demo-2 mission that transported astronauts Behnken and Hurley to the International Space Station [35].

Jared Isaacman's Inspiration4 pioneered all-civilian orbital spaceflight while raising $200 million for St. Jude Children's Hospital. His Polaris Program advances commercial space travel with scientific and humanitarian goals [38]. This evolving model has reduced space access costs and reinvigorated public interest, inspiring a new generation of space companies that complement traditional agency efforts.

6. Hope's role in inspiring future generations

As each era of space technology and mission successes unfolds, new generations of space enthusiasts emerge to carry forward advancing concepts and capabilities for long-term habitation and travel into outer space. As they see images from the Hubble and James Webb Space Telescopes, curiosity heightens, and hope for approaching those celestial bodies grows stronger.

In the previous sections, we have described resilience, curiosity, and collaboration. These are integral to hope as inspiration for addressing challenges, obstacles, and difficulties. One need look no further than the Artemis Accords to find evidence of deep-seated hope among international partnerships in space exploration [39]. Hope is expressed by those intending to travel in space, by those expecting to perform experiments and studies in space, by companies looking to find commercial opportunities in space, and by astronomers seeking knowledge of the origins and dynamics of space.

Hope extends beyond merely wishing for something and implies an expectation to obtain an objective or goal. To this end, higher education focuses on understanding the challenges and offers means to generate solutions that confront potential space travelers. Hope helps people stay calm when something less than desirable occurs and offers optimism that they will get through it. This becomes indispensable in the isolation of space [40]. Hope is motivation. It is goal-oriented. Hope exists in all cultures, as evidenced by the fantasy literature focusing on deep space travel and exploration. Current and future generations of space enthusiasts ardently seek matches for the obstacles confronting propulsion, vehicles, life support systems, and psychological dimensions of deep space travel.

Personality and psychosocial, physiological, and environmental conditions are strong catalysts for hope. Hope promotes well-being and an individual's belief in their ability to perform tasks and attain goals [41]. When unanticipated circumstances arise, which is almost a certainty in long-term space missions, hope contributes to developing a clear goal toward something better. Hopeful people believe they can influence their goals and their efforts can positively affect them. As a result, other positive states emerge, like courage and confidence. These, in turn, contribute to enhanced creativity in problem-solving. Here again, we see the influences of resilience and curiosity at play [42].

Positive expectations for the future are often conceptualized as hope. Snyder [43] explained hope as a state of positive motivation founded on three components: objectives (goals to be achieved), pathways (planning directed toward achieving those goals), and personal agency (motivation directed toward those objectives). Therein lies the foundation for inspiration to achieve necessary new materials, structures, operating systems, and support for humans and machines that will surpass existing barriers and inspire even more outstanding achievements required for safe and productive space endeavors.

The power of space to inspire hope is perhaps best illustrated through the story of Anousheh Ansari, the first female private space explorer. Anousheh Ansari exemplifies hope through her pioneering odyssey as the inaugural female private space explorer and her astute stewardship of XPRIZE, wherein she harnesses humanity's collective optimism to address our most formidable challenges [44]. Nicole Stott's Space for Art Foundation exemplifies how astronauts inspire hope through creative expression, collaborating with children worldwide on space-themed art projects [45]. In the same way, Leland Melvin's transition from NFL player to astronaut exemplifies diverse paths to space, inspiring young people from underrepresented communities to pursue their dreams boldly [46].

7. Standards and safety integration in modern space operations

Standards have been the foundation for successful enterprises in space. Human factors engineering and human-centric design featured prominently as the programs evolved, and the catalog of human factors principles applicable to space travel and habitation grew accordingly to increase safe operations and conditions. The mechanical and electrical engineering accomplishments have evolved similarly. Propulsion systems, many untried, such as solar sails and nuclear reactors, generate new types of safety issues and problems. Also, the potential for hibernation, AI, and communication, while a favorite among science fiction aficionados, is a technological area with fewer proven safety and reliability records [47].

Even with spectacular successes, notable anomalies have occurred, as evidenced in early Mercury tests, STS failed missions where engineered standards were overridden, and catastrophe resulted, or station malfunctions that affected Mir and the ISS [48]. A key consideration is that materials, procedures, operations, and habitability standards are founded on experience in the Earth's orbits or moon transits. Lessons learned from Mir and the ISS have shown that parts fail, and functional reliability is constantly threatened. Almost daily, something breaks or malfunctions on the ISS [49]. Also, the effects of neutrinos, microcosmic radiation, and other hazards are documented, although some are not well understood. On a longer transit with a spacecraft larger than any other that has moved through the solar system, the risk of collision with passing objects becomes of more significant concern [50]. Taken in the aggregate, the risks are profound. This is where resilience and hope are key components to minimize adverse behavioral consequences.

Thorough and comprehensive aspects of engineering and advanced technology, provided by experts in respective fields, have addressed the many challenges confronting extended travel in space. Physical, social, and psychosocial stressors and training principles that enhance performance have been evaluated and, to some extent, standardized, given the known measures and factors at play. Significant progress has been achieved that highlights cognition and human error, provides human reliability analyses, and addresses human-machine interfaces in space systems design [51].

The physical safety of astronauts is a subject of immense study and growing knowledge. The physiological safety for humans and machines has a substantial library of applicable evidence. Less understood are the cognitive and psychological aspects of long-term directional missions in space [52]. Suppose navigation or habitation is affected, or there are any number of situations that might threaten well-being. In that case, the resident resources of machines and humans must resolve the issues. To that end, function allocation is central, wherein the decision-making for what, how, and when resides. This requires compliance by all parties (human and machine) and execution when needed. Intuition is unreliable (as of this writing) with AI, and AI routinely questions human intuition [53]. A working resolution for such conundrums is essential.

Created in 1967 by the United Nations amid the concerns of nuclear weapons in space, the Outer Space Treaty is an international agreement governing how nations act in space. As of 2024, 115 countries were party to the agreement, with an additional 22 signatories. The key principles are to assure that exploration of outer space benefits all mankind, with no claim of sovereignty and no nuclear weapons, and that celestial bodies will be used exclusively for peaceful purposes. Other aspects call for liability for damage caused by space objects and for avoiding harmful contamination. In addition to the Artemis Accords, other treaties have built upon the Outer Space Treaty to address commercial space activities, mining, and other enterprises. These include the Rescue Agreement of 1969, the Liability Convention of 1972, the Registration Convention of 1975, and the Moon Agreement of 1984 [54].

The Artemis Accords were adopted in 2020 by the United States with 13 provisions that established a principled framework for sustainable human exploration among celestial bodies, including exploiting their natural resources. It is rooted in the provisions of the Outer Space Treaty and expands to address adaptive governance for permanent outposts and orbital stations. The Accords have developed areas of mutual interest, including peaceful purposes, interoperability of systems, transparency, emergency assistance, registration of space objects, deconfliction of space activities, space resource extraction, orbital debris, and release of scientific data. Currently, there are 54 nations that are signatories [55]. The Accords provide a strong foundation for safe practices and accountability.

The AI CRM, described earlier, is one approach to address situations unique to long-range space travel. Another concern is that behavioral markers are absent for long-duration missions. Consequently, central control becomes less influential as distance from Earth increases. A further consideration is that gravitational fields may affect cognition or physiology. The list of potential influences and hazards is extensive. To engineer standards suitable for ameliorating these conditions may only be possible once the mission has launched. Our understanding of ergonomics and human-centered design is formidable based on decades of experience in microgravity and zero-atmosphere conditions. Still, we cannot be certain that current standards will accommodate human physiological changes due to extended exposure to microgravity. Workstations, interfaces, fittings, and other adjustments would need to be possible. Nonetheless, our smaller spacecraft that have journeyed to Mars and beyond have been limited in the information obtained regarding the effects on humans in their transits [56].

Attention to psychological issues warrants increased consideration for future deep space travel. A Behavioral Health Team works with astronauts to increase the likelihood of salutogenic experiences. Salutogenesis is a process that achieves positive results by adapting to a harsh and stressful environment. This concept, which was identified and adapted for human conditions in long-duration space missions, was originally concerned with relationships between health, stress, and coping. NASA operates a Human Exploration Research Analog (HERA) habitat to simulate several missions. While not in microgravity, HERA replicates the isolation and confinement on space missions [57]. It is, together with the actual missions, the extent of our understanding regarding the primary psychological issues.

The astronaut selection process may require a greater focus on psychiatric resilience, taking into account family traumas, addictions, and personality traits associated with possible emergencies in the context of space community security [58]. Psychological screening in multiple dimensions is performed with astronauts to assess their mental suitability for space travel [59]. Similarly, there is a focus on intercultural relations and relevant psychological support. The primary basis for success is in parity of information, where parties in a transaction have equal information [60]. The selection process may become even more refined as actual missions are scheduled.

Psychological safety is a particular challenge in that, on long-duration missions, there is no return or escape alternative. In the Mir and ISS missions, the astronauts had a timeframe for their return to nearby Earth. Even the moon missions were only a few days on a round-trip. While various emergencies and unanticipated circumstances arose, the astronauts communicated with their Earth counterparts and support teams. This would not be the case for extended trips far beyond Earth. On that note, it is useful to bring to bear points raised in previous sections regarding astronaut skill evolution and lessons learned, resilience in extreme environments, CRM/TEM skills, curiosity and adaptation, benefits from prior collaborations, and the role of hope. These fundamentals and foundations will provide the pilgrims in long-duration directional missions with their toolbox for success.

8. Conclusions

Space exploration exemplifies humanity's optimism, a testimony to what we can accomplish when we search collectively and venture to reach for the stars. This chapter presented substantial insights and summarized the necessary human aspects for successful space exploration, underscoring the interplay between resilience, curiosity, collaboration, and hope as foundational traits for future astronauts. We also highlighted the need for human factors engineering and human-centric design to ensure safe operations in the evolving landscape of space missions.

Conflict of interest

The authors declare no conflict of interest.

Author details

Leila Halawi*, Mark Miller and Sam Holley
Embry Riddle Aeronautical University, Daytona Beach, Florida, USA

*Address all correspondence to: halawil@erau.edu

References

[24] NASA. Artemis Plan: NASA's Lunar Exploration Program Overview. 2020. Available from: https://www.nasa.gov/wp-content/uploads/2020/12/artemis_plan-20200921.pdf

[25] NASA. Moon to Mars Objectives. 2022. Available from: https://www.nasa.gov/wp-content/uploads/2022/09/m2m-objectives-exec-summary.pdf

[26] Lagata E. As Starliner astronauts prepare for return, look back at the mission's most significant moments. USA Today. 2025. Available from: https://www.usatoday.com/story/news/nation/2025/03/10/starliner-crew-return-iss-space-station/81950078007/

[27] National Aeronautics and Space Administration. Vostok 1. NSSDCA/COSPAR ID:1961012A. 2025. Available from: https://nssdc.gsfc.nasa.gov/nmc/spacecraft/display.action?id=1961-012A

[23] NASA John H Glenn. Former NASA Astronaut. Available from: https://www.nasa.gov/people/john-glenn/

[29] Hear the Inspiring Personal Story of Dr. Story Musgrave. Targeted News Service. Washington, D.C: Targeted News Service; 2013

[30] NASA – Sally Ride (1951-2012). NASA Astronaut/First American Woman in Space. Available from: https://science.nasa.gov/people/sally-ride/

[31] NASA – Jessica U. Meir. NASA Astronaut. Available from: https://www.nasa.gov/people/jessica-u-meir/

[32] NASA Astronaut Christina Koch. Available from: https://www.nasa.gov/people/christina-koch/)

[33] Luabeya, Monica. NASA's Newest Astronauts. 2024. Available from: https://www.nasa.gov/image-article/nasas-newest-astronauts/

[34] NASA's SpaceX Crew-9. Available from: https://www.nasa.gov/mission/nasas-spacex-crew-9/

[35] Watles, Jackie. Meet the 4-Person Crew Spearheading SpaceX's Daring Polaris Dawn Mission. 2024. Available from: https://www.cnn.com/2024/08/26/science/polaris-dawn-crew-spacex-jared-isaacman/index.html

[36] Martin R. Regional economic resilience, hysteresis and recessionary shocks. Journal of Economic Geography. 2012;**12**(1):1-32. DOI: 10.1093/jeg/lbr019

[37] Bristow G, Healy A. Innovation and regional economic resilience: An exploratory analysis. The Annals of Regional Science. 2018;**60**(2):265-284

[38] Hollnagel E. Safety-II in Practice: Developing the Resilience Potentials. New York, NY, USA: Routledge; 2017

[39] Hollnagel E, Woods DD, Leveson N. Resilience Engineering: Concepts and Precepts. 1st ed. Aldershot, England: Ashgate; 2006. DOI: 10.1201/9781315605685

[40] Hollnagel, E. RAG – Resilience Analysis Grid. 2015. Available from: https://erikhollnagel.com/onewebmedia/RAG%20Outline%20V2.pdf

[41] Suedfeld P, Steel GD. The environmental psychology of capsule habitats. Annual Review of Psychology. 2000;**51**:227-253

[42] Nick K, Manzey D. Space Psychology and Psychiatry. 2nd ed. El Segundo, Calif: Microcosm Press; 2008. Web

[43] NASA Mir Space Station. 2001. Available from: https://www.nasa.gov/history/SP-4225/mir/mir.htm#:~:text=Mir%20set%20every%20record%20in,Thagard%20began%20his%20Mir%20residency

[44] Kopeć W, Pochwatko G, Kornacka M, Stawski W, Grzeszczuk M, Skorupska K, et al. Human factors in space exploration: Opportunities for international and interdisciplinary collaboration. arXiv.org. 2024:1-13. Available from: https://arxiv.org/pdf/2403.12344

[45] Marquez JJ, Landon LB, Salas E. The next giant leap for space human factors: The opportunities. Human Factors: The Journal of the Human Factors and Ergonomics Society. 2023;**65**(6):1279-1288. DOI: 10.1177/00187208231174955

[46] National Aeronautics and Space Administration (NASA). International Space Station (ISS) EVA Suit Water Intrusion. High Visibility Close Call IRIS Case Number: S–2013-199-00005. Available from: https://www.nasa.gov/wp-content/uploads/2023/07/suit_water_intrusion_mishap_investigation_report.pdf

[47] NASA Astronaut Scott Kelly. Researchers Available to Discuss One-Year Mission. 2016. Available from: https://www.nasa.gov/news-release/nasa-astronaut-scott-kelly-researchers-available-to-discuss-one-year-mission/

[48] Miller M, Holley S. Deficits in cognitive resilience of commercial pilots: The case for adding a computer information and automation tenet in digital flight operations. Human-Centered Aerospace Systems and Sustainability Applications. 2023;**98**:74-84. DOI: 10.54941/ahfe1003921

[49] US Navy: CRM 6th Generation Briefing on Threat and Error Management, C-050-1503A PPT 5001 01-Property of US Navy (2004)

[50] Bridges KE. Human and artificial intelligence considerations for long duration space travel – A human factors perspective. Journal of Space Safety Engineering. 2023;**10**(4):433-438. DOI: 10.1016/j.jsse.2023.10.003

[51] Johnson M, Vera AH. No AI is an island: The case for teaming intelligence. AI Magazine. 2019;**40**(1):16-28

[52] 2023 space missions: further progress towards human conquest of space. CE Noticias Financieras 2023

[53] The space missions of 2023: the year of the transition to the Extraterrestrial era. CE Noticias Financieras 2023

[54] Margolis, Emily. National Air and Space Museum, Smithsonian Kalpana Chawla: The First South Asian American Woman in Space. 2023. Available from: https://airandspace.si.edu/stories/editorial/kalpana-chawla

[55] NASA Hosts Astronaut Don Pettit Prelaunch Interviews. PR Newswire. New York: PR Newswire Association LLC; 2024

[56] Columbia University Press Blog. Dr. Story Musgrave: A Surgeon in Space. 2019. Available from: https://cupblog.org/2019/05/22/dr-story-musgrave-a-surgeon-in-space

[57] Tatarewicz JN. Chapter 16 The Hubble Space Telescope Servicing Mission. Available from: https://www.nasa.gov/history/SP-4219/Chapter16.html

[58] NASA. NASA Astronauts Launch from America in Historic Test Flight of SpaceX Crew Dragon. 2020. Available from: https://www.nasa.gov/

news-release/nasa-astronauts-launch-from-america-in-historic-test-flight-of-spacex-crew-dragon/

[59] Hickok K, Howell E. Apollo 13: Facts about NASA's Near-Disaster Moon Mission. Space.com. 2023. Available from: https://www.space.com/17250-apollo-13-facts.html

[60] Pandey N. The Story of Sergei Krikalev, Whose Country Disappeared When He was in Space. 2024. Available from: https://www.ndtv.com/science/sergei-krikalev-the-story-of-sergei-krikalev-whose-country-disappeared-when-he-was-in-space-5414875

[61] Isaacman J et al. Inspiration4: Medical research advances from the first all-civilian orbital spaceflight. Aerospace Medicine and Human Performance. 2022;**93**(4):385-392

[62] Scioli A. The psychology of hope: A diagnostic and prescriptive account. In: van den Heuvel SC, editor. Historical and Multidisciplinary Perspectives on Hope. Cham: Springer; 2020. DOI: 10.1007/978-3-030-46489-9_8

[63] Bernstein G. The power of hope: Embracing resilience, inspiring change. CommunityRec. 2023. Available from: https://communityrecmag.com/the-power-of-hope-embracing-resilience-inspiring-change/

[64] Laranjeira C, Querido A. Hope and optimism as an opportunity to improve the "positive mental health" demand. Frontiers in Psychology. 2022;**13**:827320. DOI: 10.3389/fpsyg.2022.827320

[65] Campbell P. Why hope matters. Psychology Today. 2021. Available from: https://www.psychologytoday.com/us/blog/imperfect-spirituality/201902/why-hope-matters

[66] Snyder CR. Hope theory: Rainbows in the mind. Psychological Inquiry. 2002;**13**(4):249-275

[67] Ansari A. XPrize. Available from: https://www.xprize.org/about/people/anousheh-ansari

[68] Stott N. Space for Art. Available from: https://www.npsdiscovery.com/sciart

[69] Howell E. NASA Astronaut Leland Melvin talks Black History, Education, and Dogs (exclusive). 2023. Available from: https://www.space.com/nasa-astronaut-leland-melvin-black-history-education-dogs-course

[70] Johnson L, Roy K. Interstellar Travel: Propulsion, Life Support, Communications, and the Long Journey. 1st ed. San Diego: Elsevier; 2024. Print

[71] Skowran G. Significant Incidents and Close Calls in Human Space Flight. National Aeronautics and Space Administration.; 2024. Available from: https://sma.nasa.gov/SignificantIncidents/

[72] Garrett S. International Space Station: Operating an Outpost in New Frontier. National Aeronautics and Space Administration (NASA). 2018. Available from: https://www.nasa.gov/wp-content/uploads/2018/04/iss-operating_an_outpost-tagged.pdf?emrc=633ea0

[73] Shea G. Why Go to Space? National Aeronautics and Space Administration; 2023. Available from: https://www.nasa.gov/humans-in-space/why-go-to-space/

[74] Sgobba T. International Association for the Advancement of Space Safety. In: Sgobba T, editor. Space Safety and Human Performance. Oxford, England: Butterworth-Heinemann; 2018

[75] la Torre D, Gabriel G. Cognitive neuroscience in space. Life (Basel, Switzerland). 2014;**4**(3):281-294

[76] Nørgaard K, Cyborgs L-VM. Neuroweapons, and network command. Scandinavian Journal of Military Studies. 2021;**4**(1):94-107. DOI: 10.31374/sjms.86

[77] Howells K. What is the Outer Space Treaty. The Planetary Society; 2024. Available from: https://www.planetary.org/articles/what-is-the-outer-space-treaty?gad_source=1&gclid=EAIaIQobChMI58Hw067LjAMVaUH_AR2sQDrDEAAYASAAEgL39PD_BwE

[78] Deplano R. The Artemis Accords: Evolution or revolution in international space law? The International and Comparative Law Quarterly. 2021;**70**(3):799-819

[79] Szocik K, editor. The Human Factor in a Mission to Mars: An Interdisciplinary Approach. 1st ed. Cham: Springer International Publishing; 2019

[80] Lindström B, Eriksson M. Salutogenesis. Journal of Epidemiology and Community Health. 2005;**59**(6):440

[81] Kim DW. Astronaut selection and potential risk management: Psychological trauma and resilience for the Mars space mission. Trames Trames (Tallinn). 2024;**28**(3):211-236

[82] Turner DH. How space travel affects astronaut mental health. Angles. 2019. Available from: https://cmsw.mit.edu/angles/2019/headspace-how-space-travel-affects-astronaut-mental-health/

[83] Feichtinger E, Gushin V, Ryumin O, Vinokhodova A. Long-duration isolation experiments in preparation of interplanetary missions: Psychological support of international crews. Journal of Space Safety Engineering. 2022;**9**(2):223-230

Chapter 3

Perspective Chapter: Astronauts as STEM Role Models and Science Ambassadors

Lisa McNamee

Abstract

Astronauts are the most visible STEM ambassadors available to space science globally. Traditionally the astronaut role has been synonymous with the pinnacle of scientific achievement and this has translated into successful astronaut led educational initiatives worldwide. In recent years, there have been innovative approaches to leverage the astronaut role model profile further to engage all sections of society in scientific education. This has been facilitated through a variety of online, in-person, and hybrid initiatives to generate enthusiasm amongst learners for space and broader science education. This chapter explores some of the more successful established and emerging approaches to leveraging the astronaut role in STEM education.

Keywords: STEM, role model, astronaut, space, science education

1. Introduction

Astronauts are the highest profile Science Technology Engineering and Mathematics (STEM) ambassadors science outreach could ask for, motivating children and students of all ages to pursue science careers to the benefit of terrestrial innovation and health. The dedication to advanced degrees and scientific research personified by many astronauts offers a trusted STEM role model to society. This is of particular importance in a time of limited science role models and diminishing trust in the scientific community. Astronauts' broad and interdisciplinary training encourages collaboration between disparate scientific disciplines and inspires innovation at these intersections. This chapter explores how the astronaut role as science ambassador can be optimised for the benefit of terrestrial sciences.

Astronauts are recognised internationally as high profile STEM role models. At a time of intensifying demand for STEM graduates, the astronaut role may be an under leveraged tool for increasing the interest in STEM careers to a broader population cohort. In the US alone, there is a projected increase in STEM opportunities of 3 million by 2026 but an anticipated deficit of 1.5 million STEM graduates required to fill these positions. This is a pattern replicated elsewhere [1, 2]. U.S. educational initiatives have not generated significant improvements in mathematics or science literacy in the same way that European countries such as Germany have improved their outcomes in these educational metrics since 2000 [3].

An astronaut exemplifies scientific excellence in a way that can motivate children and young people to aspire to STEM careers. However, this perceived excellence of attainment may enforce the "not for me" attitude towards science that can occur in those who do not have a high level of scientific literacy at home. It is crucial that STEM engagement strategies focus on accessibility to a wide cohort of society in order to maximise impact. Inspiring those children and students who do not have other trusted science role models is of particular value in widening the appeal of STEM beyond traditionally enthusiastic cohorts. Socio-economic barriers have been identified as a difficulty in expanding the STEM workforce [3]. For women, STEM participation has been shown to particularly decrease between the ages of 11–15, with gender stereotypes, lack of role models and cultural pressures all showing an impact in disengagement [4, 5]. Engaging students in space related educational activities has been demonstrated to build their confidence in scientific knowledge through engagement with appropriate professional mentors [3].

The engagement of educators is no less important, and both NASA and ESA have programmes targeting educators at different levels. However, many educators as well as parents may see advanced STEM and space research as a luxury extra rather than a necessity. Numerous educators do not realise how many terrestrial services and technologies have developed from or are currently dependent on the space programme [6].

2. Managing a time limited resource

Astronauts are an extremely time limited resource as there are a tiny number of active career astronauts globally at any one time. Although small numbers of astronauts have travelled to space, approximately 1425 individuals up to 2023, they can have an outsized educational impact due to their public profile [7]. To date, over 43 million students worldwide have participated in educational activities related to the International Space Station (ISS) [8]. Public and student access has been facilitated through dedicated astronaut facilitated projects such as Tomatosphere, MAI-75, Sally Ride Earth Knowledge-based Acquired by Middle Schools (EarthKAM), and YouTube Space Lab [8].

Their time is further restricted by their onerous training schedules as well as the competing demands of mission training, physical training, public outreach, educational outreach and personal and family time. Therefore, all STEM and outreach initiatives utilising astronauts as a resource should ensure that maximum usage of the available time is prioritised. There have been many different strategies successfully used previously to generate maximum exposure or return from an initiative involving astronauts. Mission X, an international, multi-year educational project utilises minimal individual astronaut time to launch and generate interest in the project, while delivering most of the educational content without much required input or time commitment from the astronaut corps.

In 2011, the Mission X: Train like an Astronaut challenge was developed to promote healthy eating and physical activity to children by teaching students about the requirements of the astronaut lifestyle and encouraging them to participate in activities [9].

The challenge is a collaboration between ESA, the UK Space Agency and the European Space Education and Resource Office, housing educational activities for students where they can complete physical challenges, learn about nutrition through practical exercises and focus on their teamwork skills. Schools formed challenge teams and participated in an international fitness challenge. Participants could see the progress of teams in other countries adding an international dimension to the competition and a points system for challenges completed was utilised to promote constructive competition. A total of 4164 students participated in the initial competition challenge and the activities on the website were made available in seven languages [9].

The fitness challenge moved from strength to strength with the 2015 iteration hosting almost 40,000 children on 800 teams, from 28 countries affiliated with 12 space agencies [9].

Mission X is supported by multiple space agencies including the NASA, ESA and JAXA astronaut corps. In the 2015 iteration, they participated in the opening, closing and special ISS events. Italian astronaut Samantha Cristoferetti was the astronaut ambassador for the 2015 programme. UK astronaut Tim Peake and US astronaut Kate Rubins have been the mission X ambassadors in 2016 and 2017 [9]. In addition to participation in the opening and closing events, astronauts provided short, pre-recorded videos accessible to students, giving the astronaut's perspective on health and fitness in space. Samantha Cristoferetti was able to provide video content to students from the ISS during her space mission.

NASA and ESA offer targeted training to teachers on space science topics through online resources and focused updates courses that include classroom activities. There is a self-selection bias however in that generally only teachers who are already interested in space will seek out these opportunities and subsequently integrate them into classroom teaching. ESA's European Space Education Resource Office (ESERO) is a targeted support for early years, primary and secondary school teaching about space. It provides resources, teaching and initiatives for teachers across the continent to use space in the classroom. Each ESA member state has its own local office who will support in-country STEM promotions and space initiatives, linking students with role models in the space industry, including where possible, astronauts.

There have already been many successful efforts to incorporate astronauts into national educational curricula internationally (Netherlands, 2011; UK, 2015) to teach fundamental science skills and this important work can be further built on [10]. Incorporating space science into a national curriculum often coincides with the selection of a national astronaut, as in the case of the UK's Tim Peake. During his 2015–2016 Principia mission there was a renewed focus in UK education on space and STEM subjects [11]. During this mission, a mixed methods study examined the attitudes of 555 primary and 796 secondary school students in relation to STEM and human spaceflight. Students were found to have more positive views about space science than STEM subjects generally, with primary school students having more positive views of STEM than secondary school students. Boys generally were found to be more positive about spaceflight than girls, as well as more likely to see themselves working in the space industry in the future [11]. Over half of students felt they enjoyed learning about space generally, and felt it was important to send humans into space. Only a quarter felt they would like to have a job in the space industry with potential barriers highlighted including a need to be clever, with space science being seen as harder than maths, the next highest ranking subject [11]. The study highlighted the need for more information for students on the different types of roles available in the space sector as well as the requirement for a wide variety of role models in STEM careers [11].

Many STEM initiatives aimed at secondary schools have wholly or partially attributed their success to the patronage and support of astronauts. From 2015 to 2018, Italy ran the STEM programme, "Space for Your Future. The ISS: Innovatio, Scientia,

Sapientia" in partnership with the Italian Space Agency [12]. This was a competition aimed at secondary school students and was a successful STEM initiative involving many schools and spearheaded by astronaut Lt. Col. Walter Villadei under the School to Space research programme. The competitions thematic areas covered science, engineering and health including a "Train like an astronaut" strand which sought proposals for the development of physical exercise, tools and sensors for astronauts. The response surpassed expectations with 110 proposals submitted to the programme by 54 Italian secondary schools.

3. Astronaut public engagement

Public engagement has formed a significant part of the demands on astronaut scheduling since the inception of the role. Astronauts have engaged with this aspect of their work with varying degrees of enthusiasm. The required skills and background for astronauts has evolved over time, in part to reflect the increased requirement to engage with the public, in both science communication and in relation to STEM education [7]. This has encouraged the recruitment of individuals who are comfortable fielding media queries and confidence in public speaking [7]. With an increased level of exposure through the 24 hours news cycle and social media omnipresence, it is more crucial than ever for astronauts to engage with the public effectively.

We live in an age of online disinformation where scientific facts are regularly contested without factual basis. Easy dissemination of unscientific material via social media networks and unscrupulous individuals leads to a proliferation of online disinformation. It has been shown that increasing the science literacy of a country's population makes them less susceptible to online disinformation. Astronauts as a trusted scientific role model can aid the proliferation of good quality scientifically accurate information, while educating the public in the scientific method and the principles of hypothesis, testing and uncertainty.

Many retired members of NASA's astronaut core have continued to dedicate a significant amount of effort to promoting space, science and discovery to the public e.g., Col. Chris Hadfield, Dr. Mae Jemison, Daniel Tani. This has taken the form of school visits, public lectures and science events. Dr. Jemison, the first African-American woman in space, has continued extensive outreach campaigns focused on bringing space education to those who might not otherwise be able to access it [13]. Dr. Jemison credits seeing female NASA astronaut Dr. Sally Ride travel to space as inspiring her to apply to the astronaut corps [13]. It has been repeatedly shown that the visibility of female astronauts encourages young women to see a STEM career as more appealing [14–17].

At the advent of the space programme, U.S. astronauts were all male and marketed as "all American boys" or paragons of idealised masculinity, making them role models for American men specifically [18]. The astronaut profile was designed to role model heroism, conformity and stability [7]. In the first 20 years of spaceflight, only 2 women flew to space, while at the same time more than 182 men travelled to space [7]. In the US, Congress passed the Equal Opportunity Act of 1972 to allow women access to careers that previously were not available to them. However, NASA was hiring women prior to the federal law change [19].

Women make up 50% of the U.S. population but after they were accepted into the astronaut core, represented <25% of selected astronauts. Maintaining excellence in the competitive domain of spaceflight, requires getting the best ideas and perspectives, regardless of the background of the individuals involved. Diversity of thought and experience is necessary to promote effective scientific progress. NASA recognised this need and adopted various approaches to promote its career opportunities to women and minorities. As part of one awareness campaign, Star Trek actress Nichelle Nichols "Uhura" worked with NASA to produce a promotional video to inspire women and ethnic minorities to join NASA and to apply to the astronaut corps [7, 20]. In 2013, there were equal numbers of men and women accepted by NASA for the first time. The Canadian Space Agency has to date selected 3 women out of 14 astronauts [7]. International astronauts who have been sent to NASA have increased the diversity representation in terms of ethnicity but not gender as the majority of international astronauts that have been sent to date have been men [7].

More broadly, female participation in space missions internationally has always lagged significantly behind male participation. A study examining the number of women scientists in 10 ESA solar system missions over a 38 year period showed the percentage of women as co-investigators is always less than 16%. This is lower than the 24% of women in the International Astronomical Union during this time from ESA member states giving an estimate of the female members active in the field [21]. Women constituted 22% of the space industry workforce overall in 2021 which shows an ongoing deficit in attracting women into this sector [22]. Three Western European women have been to space to date, Helen Sharman, Dr. Claudie André-Deshays and Samantha Cristoforetti, representing 6% of astronauts of European nationality [7].

Following many years of increased efforts to promote female and ethnic minority participation in traditionally underrepresented STEM fields, there has recently been some global backsliding on these initiatives. It is therefore crucial that the success of the highest profile members of these areas are celebrated and shown to the generations coming behind to encourage a broad pipeline of STEM graduates with diverse perspectives. With the removal of celebratory pages of women at NASA, as well as sustained political attacks on diversity and inclusion measures worldwide, the real accomplishments of female astronauts must be promoted and celebrated to ensure that there remains a pipeline of motivated students to pursue STEM careers. This is in keeping with the goals of the 2030 Agenda for Sustainable Development of the United Nations, especially Goals 4 and 5 which aim to "ensure inclusive and equitable quality education and promote lifelong learning opportunities for all" and to "achieve gender equality and empower all women and girls" [23]. Social identity informed research has shown that perceptions of who is part of the in group or out group in STEM fields can be changed through educational programmes and intervention [24].

For emerging space nations, having a country's first astronaut is usually a tremendous source of pride and the individual becomes a national role model. In countries where there have been small numbers of astronauts, their selection, training and flights to ISS are generally featured on national news and attract significant political and public commentary. During British astronaut Tim Peake's Principia mission to the ISS, in addition to the multitude of ESA science experiments he took part in, also focused on educational outreach, engaging 1.6 million European children in over 30 science projects through their schools [25].

Astronauts are considered to embody key national values and to be a formal representative of their nation at high profile international events. The iconography of the astronaut was utilised as the "embodiment of national virtues" according to Launius from the Apollo era onwards [26]. A study looking at the universal values of astronauts found a general similarity in the endorsed values by individual astronauts. However, the study authors found that Canadian astronauts specifically endorsed patriotism as one of their top three values, as well as the value of universalism more than astronauts from other nations. They also appeared to value conformity less than astronaut peers from other nations [27]. This was felt to reflect Canadian cultural values and pride in their small group of astronauts [27].

The US military has utilised their military NASA astronauts who have an additional designation as space force guardians in their STEM outreach through motivational talks and the sharing of expertise, even from the ISS itself. In 2024, Col. Nick Hague spoke to STEM students from the Denver School of Science and Technology directly from ISS as part of the STEMtoSpace educational outreach programme [28].

At a time when sanctions have been applied by western states to Russia following its invasion of Ukraine, it often appears that the space domain appears exempt as bilateral agreements are still in place with Roscosmos, the Russian Space Agency. Russian cosmonauts are still living and working aboard the ISS with their NASA, ESA, CSA and JAXA colleagues and although their role is often downplayed in press releases, it is one of very few areas of public life where this cooperation with Russia still occurs at the international level.

4. Considering commercial astronauts

This expansion of types of astronaut has continued with the proliferation of astronaut roles into the commercial sphere. The exacting requirements of the term astronaut have led to some contested discussion of whose role meets the requirements of the term. This may be seen in the context of the heroism traditionally associated with the astronaut role and whether an individual has met the qualifying criteria for this standard. The space explorer as hero has traditionally been measured in terms of their achievement and contribution [29]. Astronauts are exemplars of phronesis, the Greek term meaning practical wisdom, or intelligence gleaned from and associated with practical action [29]. Whether a category of astronaut meets this threshold of phronesis needs to take account of the threshold of the core successes of heroes which include (1) attempting to scale new heights of value (2) overcoming a difficult challenge (3) risking one's life in the attempt and (4) expressing the virtue of phronesis [29]. With the advent of private space missions and space tourism, the demographics of spaceflight have dramatically changed [7]. This has an effect on the STEM role model types available to students, however there has recently been a notable backlash in some instances that may discourage participation in commercial spaceflight.

Can commercial astronauts fulfil the same STEM role model purpose as career agency astronauts? There are currently a variety of suborbital flight operators (Blue Origin, Virgin Galactic) facilitating commercial, research and mixed flights. Recently, there was significant media backlash to an all female flight that included celebrity Katy Perry, but also included aerospace engineer Aisha Bowe and Nobel prize nominee Amanda Nguyen. The mission was extensively criticised in the press for the resources used to send the flyers into space and the marketing of the flight as a feminist event [30]. Previously mixed gender commercial suborbital flights have not attracted the same public opprobrium. Many of the suborbital flights currently being flown are exclusively commercial, with participants paying for a unique travel experience, not without its own inherent risks. However, there have been several commercial suborbital flights which have also contributed to the scientific community by conducting research midflight. In 2025, the Fram-2 mission undertook 22 life-science focused experiments, including conducting the first X-ray in space [31, 32]

The 100th woman in space, Emily Calandrelli, an MIT engineer turned commercial astronaut and science communicator has engaged audiences with STEM through a New York Times' bestselling book and popular children's science shows on Netflix and Youtube.

Commercial flights may inspire a positive response from the public, such as in the case of Wally Funk who completed a suborbital flight in 2021, becoming the oldest woman to have flown to space. Funk was part of a small group of women known as the Mercury 13 who underwent some of the same testing that the male Mercury astronauts went through. It was undertaken through a private contractor to NASA from 1960 to 1961 and was designed to test the ability of women to become astronauts. Despite many of the women scoring highly, the testing was cancelled and NASA did not pursue this initiative. Funk's story of the "longest wait" to get to space resonated with a large community and her story has inspired a film, memoir, extensive media coverage and engaged diverse audiences [33]. Funk's longevity in pursuing her dream, as well as her emphasis on lifelong learning and continued pursuit of aviation despite the many hurdles put before her inspired many.

Canadian commercial astronaut and medical doctor Dr. Shawna Pandya, who is due to fly with Virgin Galactic and the International Institute for Astronautical Sciences (IIAS) has spent many years conducting STEM outreach online and via in-person events. Dr. Pandya explicitly attributes the career visibility of Dr. Roberta Bondar, Canadian Space Agency astronaut, as the inspiration for her own career trajectory. As a new generation of digital native astronauts engage online audiences with science positive role models, we can expect to see further interest in and engagement with STEM education from these cohorts.

5. Novel approaches to space and STEM education

Didactic delivery of scientific education has been the model relied upon to date in most educational contexts. There have been moves within the space community to diversify methods of science communication. ESA champion in education, Dr. Niamh Shaw, has recently finished a pilot "Town Scientist" project in the Republic of Ireland that aimed to foster STEM engagement amongst adults in marginalised communities through informal learning sessions. The informal STEM learning programme was delivered over a 10 weeks initial period aimed at adult participants. Town Scientist was designed in response to findings from the science capital model, which suggests that many adults from marginalised backgrounds lack STEM engagement opportunities, often due to systemic barriers such as educational disadvantage, social perception, and limited exposure to science [34].

Dr. Shaw partnered with trusted local institutions including the Louth County library to reduce barriers to participation in the programme. The programme's evaluation measured changes in confidence, knowledge retention, and willingness to engage in STEM discussions. At the project's outset, only 38% of participants felt confident discussing science. Many expressed feeling disconnected from STEM. By the final evaluation, 75% felt "more confident" or "very confident" engaging in STEM conversations, and 85% said they would attend future science-related events. This aligns with research highlighting the importance of informal, relatable STEM experiences in building self efficacy [35, 36]. Fostering STEM engagement has further been shown to be a robust deterrent to the absorption of disinformation and particularly of online disruptive material [37, 38].

There is currently under-utilisation of social media educational outreach by astronauts worldwide. Although social media channels can become echo chambers of misinformation and disinformation, there is significant scope to promote good quality evidence-based data via this medium, particularly to younger people. Social media offers astronauts the power to connect directly with the public, humanising space exploration and STEM subjects in an approachable manner. This increased accessibility can be effective in promoting identification with STEM role models for young people and even challenging stereotypes of STEM professionals [39].

Analysis of the Facebook posts of astronauts found that female astronauts promoted women's participation in STEM fields more frequently than male astronauts, posting on this topic once for every 10 public messages, whereas male astronauts would tackle this issue approximately one third less. However, both male and female astronauts were found to post publicly about inclusive scientific endeavours and reducing inequalities in science regularly [40].

Dr. Don Pettit, a currently active NASA astronaut, is one of the more effective space science communicators online. He has completed four spaceflights and returned to earth most recently April 19th 2025 having celebrated his 70th birthday in space. Dr. Pettit enjoys conducting his own experiments during spaceflight, that he dubs "Science of Opportunity" in addition to the mandated scientific workload he undertakes as part of his astronaut duties. These are communicated to the general public in a variety of ways, including a popular video series "Science off the Sphere" he has created showing various microgravity phenomena in an approachable manner, currently available on Youtube [41].

In an era of climate breakdown, it is essential that the scientific and social benefits of a flight outweigh any pollution emitted by the launch. One of the criticisms levelled at suborbital non research flights, carrying space tourists, is related to the significant environmental impact of the activities of this small group of tourists [42]. Observing unnecessarily high carbon activity has been shown to be demotivating in terms of climate activism to wider society as it reduces a sense of collective effort amongst the populace [42]. Many astronauts have become climate advocates, citing the overview effect experience they had during their spaceflights, and seeing part of their role as communicating the fragility of life on earth.

6. Conclusion

Astronauts as trusted role models can create a trusted scientist role model bond that increases the accessibility and inherent trustworthiness of scientific content. Following the pandemic, trust in the scientific community has been subject to disinformation at scale, both online and from political quarters. This has led to a reduction of trust in science from many quarters of society. One method of increasing the trust in science is via trusted voices such as teachers, doctors and astronauts. Astronauts have the ability to influence policy makers at national and international level to support scientific initiatives. Politicians can see a national astronaut as a valuable astronaut and role model, leading potentially to increased support for scientific endeavour.

Inventive and targeted strategies to engage the public and encourage students into STEM are needed to continue to encourage young people into STEM careers. Astronauts' time is valuable and extremely limited, therefore focused clips or excerpts of astronauts advocating for particular initiatives can be used to maximise their efficacy as has been successfully done with previous programmes including Mission X.

The Science off the Sphere Youtube series shows how physics phenomena in microgravity can be humanised and made interesting to a broad audience. If such initiatives can be broadly incorporated into science curricula, this can make space science "real" to children through effective astronaut storytelling and begin the appeal of STEM careers early on.

With the rollbacks in many countries of initiatives promoting female and diverse participation in STEM higher education and traditionally male dominated fields, there is a risk of discouraging women and minorities from programmes and careers where they may not feel naturally welcome. Given the dearth of STEM workers internationally, countries who do not encourage broad types of individuals to pursue STEM careers will have fewer candidates to choose from in advanced scientific fields to the detriment of their research and economic outputs.

In the era of climate breakdown, it is crucial that all spaceflight participants and agencies are maximising the research and technology benefits of space for terrestrial usage or a significant public backlash will see spaceflight opportunities diminished.

Author details

Lisa McNamee
Royal College of Surgeons in Ireland, Dublin, Ireland

*Address all correspondence to: mcnameelisa@rcsi.com

References

[84] Emerson Survey: 2 in 5 Americans Believe the STEM Worker Shortage is at Crisis Levels. Community Involvement | Emerson US [Online]. Available from: https://www.emerson.com/en-us/news/corporate/2018-stem-survey

[85] Reiss MJ, Mujtaba T. Should we embed careers education in STEM lessons? The Curriculum Journal. 2017;**28**(1):137-150. DOI: 10.1080/09585176.2016.1261718

[86] Morrison M. Landsharks to astronauts: Towards a new STEM culture of excellence University of Notre Dame. In: Paper ID #26893 126th Annual Conference and Exposition. Tampa, Florida: American Society for Engineering Education; 2019

[87] Study: Inclusive Environment Key to Closing STEM Gap. U.S. News & World Report [Online]. Available from: https://www.usnews.com/news/stem-solutions/articles/2016-11-23/study-girls-less-interested-in-stem-fields-perceived-as-masculine

[88] The Exact Age When Girls Lose Interest in Science and Math. CNNMoney [Online]. Available from: https://money.cnn.com/2017/02/28/technology/girls-math-scienceengineering/index.html

[89] Caldwell BS. Spaceflight-relevant stem education and outreach: Social goals and priorities. Acta Astronautica. 2015;**112**:174-181. ISSN 0094-5765. DOI: 10.1016/j.actaastro.2015.03.017

[90] Gres T et al. Study of the astronaut's profile evolution since 1961: What makes a good astronaut since then and how did society impact it? In: IAF Human Spaceflight Symposium (B3). Human Spaceflight Global Technical Session (9-GTS.2) 74th International Astronautical Congress. Baku, Azerbaijan; 2023. Available from: https://www.researchgate.net/profile/Tania-Gres/publication/375116701_Study_of_the_astronaut's_profile_evolution_since_1961_What_makes_a_good_astronaut_since_then_and_how_did_society_impact_it/links/6541925e3cc79d48c5bde1a0/Study-of-the-astronauts-profile-evolution-since-1961-What-makes-a-good-astronaut-since-then-and-how-did-society-impact-it.pdf

[91] Alleyne CW et al. The Next Generation: The International Space Station Education Accomplishments Johnson Space Center International Astronautical Conference September 28. Toronto, Canada; 2014. Available from: https://ntrs.nasa.gov/citations/20140012419

[92] Lloyd CW. Mission X: Train like an astronaut challenge. In: NTRS—NASA Technical Reports Server. Houston, Texas: Johnson Space Center; 2016. Available from: https://ntrs.nasa.gov/citations/20160001043

[93] Clements A, Curtis J, Jackson L, Lyons T. Using space as a context to enrich the teaching and learning of STEM subjects. School Science Review. 2014;**96**(355):45-52

[94] Bennett J, Airey J, Dunlop L, Turkenburg-van Diepen M. The impact of human spaceflight on young people's attitudes to STEM subjects. Research in Science & Technological Education. 2019;**38**(4):417-438. DOI: 10.1080/02635143.2019.1642865

[95] Bonacci E. The international space station (ISS) contest as STEM

educational project. Journal of Sciences. 2023;**10**(3):153-180

[96] Mathias M. Mae Jemison Biography, National Women's History Museum. Available from: https://www.womenshistory.org/education-resources/biographies/mae-jemison

[97] Women at NASA. Available from: https://www.nasa.gov/women-at-nasa/ [Accessed: 28 April 2025]

[98] Herrmann SD, Adelman RM, Bodford JE, Graudejus O, Okun MA, Kwan VSY. The effects of a female role model on academic performance and persistence of women in STEM courses. Basic and Applied Social Psychology. 2016;**38**(5):258-268. DOI: 10.1080/01973533.2016.1209757

[99] González-Pérez S, Mateos de Cabo R, Sáinz M. Girls in STEM: Is it a female role-model thing? Frontiers in Psychology. 2020;**11**:1-21. Available from: https://www.frontiersin.org/journals/psychology/articles/10.3389/fpsyg.2020.02204

[100] Gladstone JR, Cimpian A. Which role models are effective for which students? A systematic review and four recommendations for maximizing the effectiveness of role models in STEM. International Journal of STEM Education. 2021;**8**:59. DOI: 10.1186/s40594-021-00315-x

[101] Llinares D. The Astronaut: Cultural Mythology and Idealized Masculinity. Newcastle: Cambridge Scholars Publishing; 2011

[102] Foster A. Integrating Women into the Astronaut Corps, Politics and Logistics at NASA, 1972-2004. Baltimore: The Johns Hopkins University Press; 2011. Available from: https://muse.jhu.edu/pub/1/oa_monograph/book/11300/pdf

[103] Dunbar B. Nichelle Nichols Helped NASA Break Boundaries on Earth and in Space. Houston, Texas: National Aeronautics and Space Administration; 2022. Available from: https://www.nasa.gov/people-of-nasa/special-observances/nichelle-nichols-helped-nasa-break-boundaries-on-earth-and-in-space/

[104] Piccialli A, Rathbun JA, Levasseur-Regourd A-C, Määttänen A, Milillo A, Rengel M, et al. Participation of women scientists in ESA solar system missions: A historical trend. Advances in Geosciences. 2020;**53**:169-182. DOI: 10.5194/adgeo-53-169-2020

[105] United Nations. Only around 1 in 5 space industry workers are women. UN News. Available from: https://news.un.org/en/story/2021/10/1102082. 2021 [Accessed: 19 July 2023]

[106] United Nations, Department of Economic and Social Affairs. Sustainable Development SDG 5 Achieve Gender Equality and Empower All Women and Girls. Available from: https://sdgs.un.org/goals/goal5

[107] Kim AY, Sinatra GM, Seyranian V. Developing a STEM identity among young women: A social identity perspective. Review of Educational Research. 2018;**88**(4):589-625. DOI: 10.3102/0034654318779957

[108] European Space Agency. Tim Peake Astronaut. Available from: https://www.esa.int/Science_Exploration/Human_and_Robotic_Exploration/Astronauts/Timothy_Tim_Peake [Accessed: 28 April 2025]

[109] Launius R. Heroes in a vacuum: The Apollo astronaut as cultural icon, session: HIS-3: Missiles and spaceflight. In: 43rd AIAA Aerospace Sciences Meeting and Exhibit; 10-13

January 2005; Reno, Nevada. 2012. DOI: 10.2514/6.2005-702

[110] Brcic J, Della-Rossa I. Universal values of Canadian astronauts. Acta Astronautica. 2012;**80**:46-51. ISSN 0094-5765,. DOI: 10.1016/j.actaastro.2012.05.014

[111] Thompson J. A Cosmic Connection: USSF Astronaut Calls STEM Students from ISS. Buckley Space Force Base; 2024. Available from: https://www.buckley.spaceforce.mil/NEWS/Article/3977350/a-cosmic-connection-ussf-astronaut-calls-stem-students-from-iss/

[112] Tachibana K. Condition of heroes: Looking for the heroic figure of future astronauts. Frontiers in Space Technologies. 2024;**5**. DOI: 10.3389/frspt.2024.1337296. Available from: https://www.frontiersin.org/journals/space-technologies/articles/10.3389/frspt.2024.1337296/full

[113] O'Connor R. Katy Perry 'Regrets Public Spectacle' of Blue Origin Space Mission, Independent. 2025. Available from: https://www.independent.co.uk/arts-entertainment/music/news/katy-perry-blue-origin-memes-reaction-backlash-b2735620.html

[114] Chu J. Massachusetts Institute of Technology 3Q: MIT's Lonnie Petersen on the First Medical X-Ray Taken in Space. 2025. Available from: https://news.mit.edu/2025/3-questions-lonnie-petersen-first-medical-x-ray-taken-in-space-0407

[115] Billings L. SpaceX hits new milestone with Fram2, the first-ever crewed polar mission. Scientific American. 31 Mar 2025. Available from: https://www.scientificamerican.com/article/spacexs-fram2-mission-sends-four-private-astronauts-into-polar-orbit/

[116] ESA. Mercury 13. 2013. Available from: https://www.esa.int/About_Us/50_years_of_ESA/50_years_of_humans_in_space/Mercury_13

[117] Archer L, Dawson E, DeWitt J, Seakins A, Wong B. "Science capital": A conceptual, methodological, and empirical argument for extending bourdieusian notions of capital beyond the arts. Journal of Research in Science Teaching. 2015;**52**:922-948. DOI: 10.1002/tea.21227

[118] Bandura A. Self-Efficacy: The Exercise of Control. New York: W H Freeman/Times Books/Henry Holt & Co.; 1997

[119] Shaw N. Town Scientist Executive Summary, Research Ireland Discover Programme. Republic of Ireland: Science Foundation Ireland; 2025

[120] Guess AM, Lerner M, Lyons B, Montgomery JM, Nyhan B, Reifler J, et al. A digital media literacy intervention increases discernment between mainstream and false news in the United States and India. Proceedings of the National Academy of Sciences. 2020;**117**:15536-15545. DOI: 10.1073/pnas.192049811

[121] Cagle SM, Anderson AA, Kelp NC. Stop the spread: Empowering students to address misinformation through community-engaged, interdisciplinary science communication training. Journal of Research in Science Teaching. 2025;**62**(3):721-755. DOI: 10.1002/tea.21971. Epub 2024 Jul 16

[122] Steinke J. Cultural representations of gender and science: Portrayals of female scientists and engineers in popular films. Science Communication. 2005;**27**(1):27-63. DOI: 10.1177/1075547005278610

[123] Loner E, Fattorini E. From space to social media: How female and male NASA astronauts promote women's participation in STEM on Facebook. The Communication Review. 2025;**28**:1-23. DOI: 10.1080/10714421.2025.2482371

[124] Pettit D. Science off the Sphere YouTube Video Series. Available from: https://www.youtube.com/playlist?list=PLB7D02AAB8C908B8C [Accessed: 29 April 2025]

[125] Westlake S. Why Katy Perry's celebrity spaceflight blazed a trail for climate breakdown. The Conversation. 2025. Available from: https://theconversation.com/why-katy-perrys-celebrity-spaceflight-blazed-a-trail-for-climate-breakdown-254824

Chapter 4

The New 'Right Stuff': What It Takes to Be an Astronaut of the Future

Donna L. Roberts

Abstract

The twenty-first-century astronaut is no longer defined solely by physical endurance and technical expertise. As space exploration expands beyond government programs to include commercial, military, and even civilian participants, the qualifications for becoming an astronaut have evolved. This chapter explores how shifting mission objectives, technological advancements, and the growing role of automation are reshaping the skills, training, and psychological resilience required for space travel. It will examine astronaut selection criteria, highlighting the increasing emphasis on cognitive flexibility, problem-solving under uncertainty, and teamwork in high-stakes environments. Additionally, it will address the growing importance of diversity—across gender, culture, and expertise—in shaping a more adaptable astronaut corps. By drawing insights from aviation, deep-sea exploration, and human performance research, this chapter will define the new "Right Stuff" necessary for future missions, from suborbital flights to deep-space colonization. Ultimately, it argues that the astronauts of the future must be as mentally agile as they are physically prepared, embodying resilience, collaboration, and adaptability to push the boundaries of human spaceflight.

Keywords: human performance in space, astronaut selection, future space missions, space mission adaptability, next-generation astronauts, psychological resilience, twenty-first century space explorer, psychological profile of future astronauts

1. Introduction

The concept of the "Right Stuff" emerged during the dawn of human spaceflight, encapsulating a sense of extraordinary courage, composure, and unrelenting determination that distinguished the earliest astronauts from other aviators [1]. During the Mercury and Apollo eras, candidates were typically elite test pilots—military officers renowned for their skill at high-speed, high-risk flight and their ability to remain calm under extreme pressure [2]. This archetype reflected nationalistic ambitions, Cold War competition, and the rigorous demands of pioneering a new frontier. Physical fitness and technical prowess were paramount, as the risks associated with ascending atop volatile rockets and orbiting Earth were staggering by the standards of the time. Yet, even in those early missions, there was a dawning recognition that psychological resilience—the capacity to adapt mentally to unexpected challenges—would prove just as critical as any engineering solution or pilot skill [1].

Over the decades, the image of astronauts began to shift, reflecting broader changes in society's understanding of expertise, diversity, and the purposes of space exploration [2]. Early on, scientists, medical doctors, and eventually educators joined the once-exclusive ranks of military-trained pilots, bringing new perspectives and academic specializations into orbit. Agencies such as NASA recognized that fulfilling complex research objectives—ranging from life sciences to advanced materials experiments—necessitated a broader skill set than test piloting alone could provide [3]. The Space Shuttle era, for instance, underscored the importance of multidisciplinary teams who could not only pilot spacecraft but also conduct detailed experiments, maintain life support systems, and handle intricate docking procedures with space stations. This gradual expansion continued to redefine the "Right Stuff," orienting it more toward versatility, team cohesion, and scientific acuity.

Central to this evolution is the understanding that twenty-first-century spaceflight involves a uniquely demanding psychological component, shaped by both physical and social isolation [1]. As missions extend beyond Earth orbit—to the Moon, Mars, or potentially even deeper into the solar system—the mental fortitude required of astronauts grows exponentially. Extended confinement in small habitats, delayed communication with Earth, and the constant presence of life-threatening hazards can induce stress responses rarely encountered in conventional professions. Astronauts now face a hyper-complex environment: simultaneously preserving their own emotional equilibrium, collaborating with international crewmates who may come from vastly different cultural backgrounds, and executing critical scientific or operational tasks with minute precision [2]. This psychological dimension, once an implied component of the job, has become a field of rigorous academic study, leading to innovative methods in astronaut training and mental health support [1].

In tandem with these changing psychological demands, the profile of who might become an astronaut has diversified, reflecting a broader shift in the global space community [3]. Government agency astronauts from organizations like NASA, ESA, JAXA, and Roscosmos still undergo some of the most stringent selection processes, focusing on academic, medical, and psychological criteria in addition to flight experience. Commercial astronauts, employed by private companies such as SpaceX, Blue Origin, or Virgin Galactic, must master not only core flight competencies but also engage in public outreach, corporate branding, and technology demonstrations [4]. Space tourists, meanwhile, typically receive compressed versions of astronaut training, enabling them to enjoy short suborbital or orbital experiences, often emphasizing the excitement of microgravity and Earth observation more than extended operational duties. Furthermore, as humanity contemplates establishing a permanent presence on the Moon or Mars, the concept of a "settler astronaut" emerges—individuals who may reside for years, if not lifetimes, in off-world habitats, contending with total dependence on life-support systems and extreme environmental constraints [3].

These widening categories—professional astronauts, commercial spacefarers, tourists, and prospective settlers—call for a new articulation of the "Right Stuff" that balances technical expertise with profound psychological resilience [2]. While the ability to perform under duress remains paramount, individuals who embark on lengthy expeditions need robust coping mechanisms for isolation, a repertoire of strategies for maintaining motivation in monotonous conditions, and exceptional interpersonal skills to foster harmony in small crews [1]. Long-duration missions

DOI: http://dx.doi.org/10.5772/intechopen.1010596

particularly underscore the importance of emotional intelligence, as trivial disagreements or cultural misunderstandings can escalate under the stress of confinement, with mission-critical consequences. In such scenarios, success depends as much on empathy, conflict resolution, and cooperation as on mastery of spacecraft systems. Hence, space agencies and private enterprises alike now prioritize attributes such as emotional regulation, adaptability to dynamic situations, and collaborative problem-solving as highly as classical test pilot aptitudes [4].

Technological advances also shape the psychological landscape in which astronauts operate. Highly automated spacecraft and artificial intelligence can reduce certain manual tasks, providing real-time diagnostics or even assisting in decision-making during emergencies [3]. However, reliance on advanced systems introduces new forms of stress, such as the need to troubleshoot software errors mid-flight or rapidly adapt to shifting mission parameters dictated by algorithms. Furthermore, the shift toward commercial space stations, lunar bases, and deep-space habitats requires astronauts to assume multiple roles: pilot, scientist, engineer, diplomat, and even crisis negotiator. Each role comes with distinct psychological pressures, from ensuring scientific results remain valid under microgravity conditions, to addressing tension within a crew that might be months of travel away from any rescue [2]. These overlapping responsibilities signify that the "Right Stuff" must expand to encompass extensive cross-training and unwavering psychological readiness.

Equally pertinent is the emergence of space tourism, a domain once relegated to science fiction, but now materializing through suborbital flights and orbital stays. Tourists paying significant sums to experience a few days or weeks in microgravity often require substantial preparation, despite the limited scope of their missions [4]. They must learn basic emergency protocols, adapt to rapid acceleration, and mentally prepare for disorientation in zero-g. For many, the anticipation and excitement—coupled with the possibility of motion sickness, anxiety, or culture shock—necessitate psychological support akin to professional astronaut programs, albeit condensed. Such training aims to ensure that space tourists remain safe and psychologically composed, preventing them from endangering themselves or others in the high-stakes environment of space travel. Even these short-duration flight participants, therefore, reflect the broader trend that mental preparedness is inseparable from the new "Right Stuff."

Looking ahead to longer and more ambitious expeditions—such as a multi-year journey to Mars—the role of psychological resilience and adaptability becomes even more pronounced [3]. Crews would have to endure chronic isolation, limited communication windows, and an unrelenting awareness that any help from Earth may arrive weeks too late. Strategies for maintaining crew morale—through meaningful work, creative social activities, and robust mental health protocols—are not optional luxuries but existential necessities. Prolonged microgravity exposures, potential radiation risks, and the high potential for conflict in cramped quarters mean that mission planners must prioritize mental well-being from the earliest stages of candidate selection to the final mission timeline. In this new era, the intangible qualities of self-awareness, emotional stability, and interpersonal insight are as crucial as physical conditioning or advanced degrees in engineering [1, 2].

As the demographic of space explorers broadens, the shift in astronaut identity—from solitary test pilots to a collective of specialists, explorers, tourists, and would-be settlers—symbolizes a transformative era in human spaceflight [2]. The very notion of exploration itself has evolved: no longer purely a geopolitical race, but rather a multinational, and increasingly commercial, venture fueled by science, economic opportunities, and the intrinsic human desire to expand our horizons. Correspondingly, the demands placed upon astronauts—both psychologically and socially—will continue to grow, necessitating a reevaluation of every stage of selection, training, and in-flight support. The classic image of an astronaut as an unflappable lone hero gives way to the reality of the twenty-first-century spacefarer, a multi-talented individual or team player who can adapt not just physically but emotionally and cognitively to the radically changing environments beyond Earth [1]. Thus, the "Right Stuff" is no longer defined by raw daring or mechanical aptitude alone; it now encompasses a sophisticated blend of emotional intelligence, collaborative spirit, and unparalleled resilience, marking the dawn of a new era of space exploration.

2. The evolution of astronaut selection criteria

The path to becoming an astronaut has undergone a dramatic transformation since the inception of the U.S. human spaceflight program in the late 1950s. Early selection campaigns were narrowly focused on recruiting military test pilots who exemplified physical courage and unwavering composure, reflecting the era's conflation of daring aerial accomplishments with national prestige [1]. Over time, however, the demands of spaceflight evolved from test-piloting experimental crafts to living in orbit for extended durations, executing complex scientific missions, and eventually preparing for long-term settlement on other celestial bodies [2]. These shifting challenges have driven agencies worldwide to systematically refine and expand their criteria for astronaut qualification. No longer is physical prowess alone sufficient; modern astronaut selection processes weigh academic excellence, interpersonal skills, and psychological resilience equally with athletic capacity or flight experience [3]. In doing so, contemporary space programs have acknowledged that success hinges as much on mental adaptability and cross-disciplinary teamwork as on any single technical competence.

This broadening of astronaut selection criteria signals a pivotal change in the philosophy behind human space exploration. During the Mercury, Gemini, and Apollo programs, candidates were chosen almost exclusively from ranks of high-performance test pilots, as the earliest missions required individuals who could handle unpredictable mechanical failures and make split-second decisions in untested flight regimes [4]. Yet, with the emergence of the Space Shuttle era, larger crews and more extensive scientific experimentation necessitated including scientists, engineers, physicians, and educators. By the time of the International Space Station (ISS), the so-called "Right Stuff" had already shifted to encompass capabilities that transcended simple test-flying aptitude, emphasizing refined communication skills, cultural awareness, and the capacity to operate in tightly knit international teams [3]. This evolution continues unabated in the twenty-first century, influenced by the advent of commercial spaceflight, space tourism, and the prospect of permanent off-world settlements.

2.1 The Mercury era to the ISS era

During NASA's earliest missions, the criteria for astronaut selection embodied the Cold War imperatives of demonstrating technological and military supremacy [1]. Officials viewed decorated test pilots as emblematic of national heroism—individuals capable of not just enduring the extreme g-forces of launch but also exuding the stoic bravado required to reassure an apprehensive public [2]. These men (and initially, they were exclusively men) possessed rigorous flight test experience in supersonic jets, backgrounds in aeronautical engineering, and near-perfect physical health. Psychological adaptability was considered an inherent quality of test pilots trained to cope with life-threatening emergencies at Mach speeds [4]. Consequently, NASA's primary focus rested on attributes measurable through medical and flight records, while more nuanced qualities like emotional resilience and teamwork were viewed as secondary or simply presumed to be present by virtue of a pilot's elite status.

When NASA advanced to projects like Gemini and Apollo, the scope of missions grew more intricate, often involving prolonged stays in spacecraft and complex in-orbit tasks [3]. While pilot expertise remained central, mission planners recognized that robust interpersonal skills, scientific curiosity, and a willingness to collaborate across newly formed mission control teams could spell the difference between success and failure. Astronauts needed to handle advanced spacecraft systems, conduct extravehicular activities, and operate rudimentary laboratories in microgravity. This progression spotlighted the importance of adaptability and mental focus under alien conditions, including isolation from family, microgravity-induced disorientation, and the psychological pressure of potential catastrophic failure. By the time the Apollo missions reached the Moon, NASA had accumulated substantial knowledge about how astronauts behaved under duress, informing the next wave of selection processes [1, 4].

The Space Shuttle era epitomized a broader team-based approach, ushering in mission specialists from various scientific and engineering backgrounds who participated in orbital endeavors lasting over a week or more [2]. As tasks broadened to satellite deployment, SpaceLab experimentation, and ISS construction, the role of a single "pilot-hero" became diffuse. Instead, the shuttle typically carried a culturally and professionally diverse crew, each member entrusted with specialized responsibilities. Moreover, gender and ethnic diversity gained traction, albeit gradually, as NASA recognized that inclusivity was not only a social imperative but also a strategic advantage, tapping into a broader range of talents [5]. By the time the ISS became operational, astronauts from different national space agencies had to live and work alongside each other for months on end—an environment that further demanded flexibility, conflict resolution skills, and the ability to forge camaraderie despite linguistic and cultural differences [3]. These evolving competencies laid the groundwork for the present-day criteria that value team cohesion and psychological fortitude as highly as piloting and engineering abilities.

2.2 The expanding astronaut corps

Today, the astronaut corps spans more than just government space agencies. Commercial entities like SpaceX, Blue Origin, and Axiom Space have begun selecting, training, and flying their own astronauts—a move that has triggered a reevaluation of what it means to be "astronaut material" [3]. On the one hand, commercial operators still adhere to many NASA-established medical and educational guidelines, seeking capable individuals with advanced technical or scientific backgrounds. On the other hand, these companies must also cater to corporate objectives, public relations, and the demands of private clientele. Astronaut candidates may therefore need to be adept not only at microgravity operations but also at media engagement and outreach events that build brand awareness or secure investor confidence [2, 5]. This pivot underscores the increasingly hybrid nature of the modern astronaut role: part researcher, part explorer, part ambassador.

Furthermore, the concept of space tourism is no longer hypothetical but an accelerating reality. Billionaires and private citizens have already embarked on orbital and suborbital flights, illustrating how selection processes can stretch beyond the traditional frameworks of scientific or operational excellence [4]. While space tourists often receive condensed training protocols focusing on safety, emergency procedures, and zero-g orientation, the psychological demands remain significant. Short-term visitors may still experience acute anxiety, sensory overload, or cultural misalignments with career astronauts or fellow tourists from different backgrounds. Consequently, even these private tourists must demonstrate some capacity for emotional regulation and teamwork under potentially stressful and unfamiliar conditions [5]. As commercial missions to Earth's orbit, the lunar surface, and eventually Mars become viable, the necessity of more inclusive and flexible selection criteria will grow exponentially.

Parallel to commercial ventures, non-traditional programs—such as those that recruit educators, artists, or humanitarian workers for specialized missions—have further broadened the astronaut profile. Entities like Space for Humanity and other philanthropic or research-focused organizations strive to democratize access to space by sponsoring individuals who represent diverse socioeconomic, cultural, and professional backgrounds [3]. These programs emphasize the educational or inspirational value of space travel, selecting participants who can translate their experiences into broader benefits for society. Here again, effective communication skills, psychological adaptability, and a capacity to thrive under scrutiny become decisive factors in evaluating who will carry such missions forward [2]. The resulting astronaut corps is increasingly varied, weaving together military pilots, accomplished researchers, corporate brand ambassadors, citizen explorers, and social entrepreneurs.

2.3 New selection priorities

One of the most noteworthy changes in astronaut selection is the explicit integration of psychological assessment as a key requirement, reflecting the critical role of mental and emotional well-being in spaceflight [1]. Modern selection panels employ diverse psychometric tools, interviews, group scenarios, and stress tests to evaluate how prospective astronauts cope with uncertainty, interpersonal friction, and prolonged isolation. This approach extends beyond the rudimentary psychological evaluations of earlier eras, which aimed primarily to screen out candidates with overt issues. Instead, contemporary processes seek positive psychological traits—such as empathy, conflict resolution, and robust stress-management techniques—that enhance overall team function [3]. Moreover, attributes like situational awareness, cognitive flexibility, and the ability to remain calm under ambiguous or rapidly evolving conditions are systematically measured, ensuring that selected astronauts can adapt to the wide range of tasks required during modern missions.

Teamwork, a long-standing but sometimes underemphasized aspect of astronaut selection, has been elevated to a formal criterion, underscoring the collaborative nature of space operations [2]. Crew members must be prepared to live, work, and solve problems in close quarters with individuals from varied backgrounds. Language differences, cultural norms, and personal habits can create friction in an environment where personal space is inherently limited. Hence, the ability to communicate effectively, demonstrate respect for diversity, and maintain psychological stability despite inevitable challenges is viewed as central to astronautic success [3]. Training in group cohesion and conflict management—once secondary to spacecraft systems and flight drills—is now integral to many space agency curricula, reflecting a reality in which a breakdown in crew relations can be as disastrous as an equipment malfunction.

Additionally, mission planners prioritize leadership potential—defined not merely by rank or hierarchical control, but by the capacity to inspire, coordinate, and protect team members under extraordinary stress [4]. In deep-space missions, where real-time support from Earth is either delayed or impossible, leadership skills must be distributed across the crew to ensure decentralized decision-making. This shift necessitates a holistic assessment of candidates' capacity to step into leadership roles dynamically, and to follow effectively when another crewmate's expertise is paramount. The result is a more democratic model of leadership, one that fosters agility and mutual support rather than rigid chain-of-command structures, and is increasingly reflected in astronaut selection panels that value candidates who exhibit self-awareness and a willingness to learn from others [2].

2.4 The role of AI and automation

As robotic systems and artificial intelligence become more capable, astronauts must also contend with the implications of sharing mission responsibilities with semi-autonomous or fully autonomous systems [3]. This collaboration with advanced technology reshapes the astronaut's role from primarily executing tasks to overseeing, optimizing, and troubleshooting complex computational processes. Selection committees now look for individuals who can maintain situational awareness while coordinating with AI-driven tools, balancing trust in algorithmic outputs with the critical thinking necessary to identify errors [1]. The ability to interpret, manage, and even reprogram automated systems under duress—particularly in environments lacking immediate ground support—has emerged as a coveted skill, merging technical aptitude with emotional steadiness.

Moreover, reliance on AI may amplify certain psychological demands. Astronauts could experience heightened stress if or when an automated system behaves unpredictably, prompting them to rapidly assume manual control or corrective measures [2]. In addition, the presence of sophisticated machinery can alter group dynamics, as crew members debate the relative authority of human expertise versus AI suggestions in real time. Ensuring that crews possess not only the technical skill to manage AI but also the collaborative mindset to accept or override machine-driven instructions responsibly has become a core element of astronaut selection [4]. Ultimately, humans remain integral to spaceflight precisely because of their unique capacity to innovate, empathize, and adapt to sudden anomalies—qualities that even the most advanced AI cannot replicate to the same depth.

In light of all these considerations, astronaut selection today involves a multifaceted evaluation of physical, cognitive, emotional, and collaborative proficiencies, shaped by lessons learned from historical missions and propelled by emergent demands of deep-space travel, commercial enterprise, and automation [3, 5]. This evolution stands as a testament to the ongoing transformation of space exploration itself, where scientific inquiry, economic opportunity, and human curiosity converge in an ever-expanding theater of possibility. As the astronaut corps continues to grow in diversity and function, the complexity of selecting candidates well-suited to tackle present and future missions underscores just how far we have come from the early days of pulling top test pilots into nascent spacecraft. Now, as we strive to venture deeper into the solar system and possibly establish long-term off-world habitats, the very definition of who qualifies as an astronaut—and why—will remain in flux, shaped by the dynamic interplay of technology, humanity, and our collective aspirations.

3. The expanding role of astronauts in future space exploration

The scope of human spaceflight has broadened considerably in recent years, shifting from government-led orbital missions to an expansive array of commercial, strategic, and deep-space endeavors [1]. In this evolving landscape, astronauts serve not only as highly trained operators of complex spacecraft but also as researchers, public ambassadors, and potential settlers of extraterrestrial environments. Government agencies still hold a foundational place in human space exploration, yet the rapid growth of private companies—combined with military and strategic interests—has reshaped the pathways and purposes of sending people beyond Earth [2]. These new roles call for an even more diverse set of proficiencies, incorporating not only technical mastery and resilience but also entrepreneurial acumen, public-facing communication skills, and the ability to adapt to multi-year missions [3]. As a result, the "astronaut experience" now encompasses everything from short suborbital jaunts for space tourists to visionary plans for permanently inhabited colonies on the Moon or Mars.

3.1 Commercial spaceflight and space tourism

Over the past decade, commercial spaceflight has emerged as a major force in low Earth orbit missions, enabling private entities such as SpaceX, Blue Origin, and others to deliver both crewed and cargo services [2]. These companies, in collaboration with government agencies, have pioneered reusable rocket technologies, lowering the cost of access to space and opening new opportunities for individuals outside the traditional astronaut corps. While government space agencies retain oversight for flight safety and mission objectives, commercial carriers undertake vehicle design, mission planning, and flight operations, thus diversifying the range of hardware and human factors strategies in play [4]. In many cases, the training, selection, and in-flight roles of commercial astronauts reflect a hybrid approach: grounded in NASA's rigorous standards but also shaped by private enterprise's priorities for efficiency, innovation, and brand outreach.

A notable offshoot of commercial operations is the emergence of space tourism, which extends the astronaut experience to paying passengers, often with backgrounds far removed from aerospace or research [5]. These participants undergo condensed training programs to prepare them for the unique physiological stresses of launch and re-entry, as well as the microgravity environment. While their missions are brief relative to those of career astronauts, tourists may still face the psychological pressure of high-stakes travel and the potential for disorientation or motion sickness. From a broader perspective, space tourism not only democratizes access to low Earth orbit but also serves as a revenue stream for commercial providers, catalyzing further technological development and potentially accelerating plans for orbital hotels or lunar tourism. In this sense, the astronaut's role—particularly within private missions—encompasses guiding and supporting less experienced travelers, ensuring safety while also fostering a memorable and informative experience.

3.2 Military and strategic space operations

Human spaceflight has historically carried military undertones, especially during the Cold War era when rocket technology was closely tied to ballistic missile development [3]. In the modern context, various nations continue to maintain or explore the possibility of crewed space operations for strategic purposes, though explicit militarization of space remains both controversial and regulated by international agreements. Nevertheless, astronauts with military backgrounds contribute specialized skill sets, including disciplined operational thinking, familiarity with chain-of-command structures, and acute survival training [2]. Such attributes can prove critical for missions in orbital or cislunar environments where real-time, high-stakes decision-making is essential for protecting valuable assets or responding to emergencies.

Additionally, human presence in space might serve as both a deterrent and a diplomatic signal in regions beyond Earth's immediate orbit [3]. As nations recognize the importance of space-based resources—ranging from satellite constellations supporting communication and navigation to prospective mining ventures on celestial bodies—trained personnel may be tasked with overseeing or defending these endeavors. In scenarios where multi-national cooperation is paramount, astronauts could also function as envoys, fostering trust and collaboration in orbit. The strategic dimension of spaceflight thus elevates the astronaut's role from that of pilot or researcher to potential mediator, guardian, and caretaker of humanity's expanding off-world presence.

3.3 Deep-space exploration and colonization

Beyond the confines of low Earth orbit, astronauts now stand on the threshold of deep-space exploration, particularly as plans for lunar bases and crewed missions to Mars become increasingly concrete [1, 4]. The Artemis program, for example, envisions sustained human presence on and around the Moon, employing new spacecraft and habitat systems to gain experience essential for eventual journeys to Mars [6]. In such missions, astronauts must cope with extended durations away from Earth, increased communication delays, and dependence on advanced life-support technologies. They will likely serve in multiple capacities—pilots, scientists, engineers, medics, and psychologists—merging a broad array of technical and interpersonal competencies to ensure mission stability in a harsh, often unpredictable environment [1].

Looking further ahead, colonization prospects raise profound questions about the astronaut's identity and purpose. Once a crew embarks on a multi-year voyage to Mars or establishes a permanent settlement on the lunar surface, they shift from being short-term visitors to functioning as off-world residents [6]. The psychological, medical, and ethical implications of such missions are daunting: astronauts must maintain robust mental health in claustrophobic environments, manage physical deterioration due to partial or zero gravity, and contend with the existential reality of potentially never returning to Earth. This blurring of lines between exploration and colonization introduces significant moral and logistical complexities, as spacefarers become de facto pioneers shaping entirely new societies. In such scenarios, the astronaut's role transcends mission completion; they become cultural stewards, community-builders, and long-term custodians of humanity's reach beyond Earth.

3.4 Psychological, ethical, and logistical considerations

At every level—from commercial flights to permanent colonies—the astronaut's responsibilities now include a host of psychological and ethical considerations. The stress of continuous media scrutiny and commercial obligations, for instance, can weigh heavily on those participating in privately funded missions [4]. Meanwhile, for deep-space ventures, the psychological toll of living in microgravity or fractional gravity for months or years demands substantial preparation in stress management and interpersonal dynamics. Each crewed mission must also account for the moral duty to safeguard astronaut well-being, balancing the inherent risks of exploration with the obligation to protect human life [1]. The confluence of commercial, military, and scientific objectives intensifies these demands, necessitating clear protocols, robust training regimens, and support systems that address the unique vulnerabilities of people operating in such extreme conditions.

From a logistical perspective, the expanding role of astronauts requires ongoing improvements in habitat design, life-support infrastructure, and communication networks [2]. Missions beyond low Earth orbit must incorporate redundancy, resource recycling, and contingency planning on an unprecedented scale. Astronauts will inevitably become specialists in situ resource utilization, tasked with converting local materials into fuel, water, or building supplies. At the same time, human adaptability remains irreplaceable: even as AI and robotics advance, the capacity for creative problem-solving, empathy, and intuitive decision-making remains a uniquely human contribution to space exploration [5]. Consequently, it is the synergy between humans and technology, rather than a simple handover of tasks to machines, that will shape the future frontiers of crewed missions.

Clearly, the astronaut's role in modern space exploration has evolved far beyond the initial archetype of a solitary test pilot pushing the boundaries of high-altitude flight [2, 3]. Today's spacefarer must balance the demands of commercial viability, strategic imperatives, and scientific innovation, all while exhibiting extraordinary psychological resilience and collaborative finesse. As humanity's ventures into space deepen and diversify—from quick orbital flights to the monumental goal of permanent off-world bases—the range of astronaut responsibilities continues to widen. Understanding and preparing for these expanded roles is essential, not only for the success of individual missions but also for the broader narrative of how humanity establishes a sustained presence among the stars.

4. Essential skills for the twenty-first-century astronaut

The demands of human space exploration in the twenty-first century extend well beyond the boundaries of piloting expertise or raw physical endurance. As missions grow more complex—whether in low Earth orbit, on the Moon, or during prospective journeys to Mars—astronauts must possess a holistic set of skills that fuse technical competence with mental resilience, cross-disciplinary know-how, and advanced interpersonal capabilities [1]. These modern demands spring from the recognition that exploration no longer revolves solely around operating spacecraft; it encompasses navigating emotionally charged conditions such as extended confinement, unprecedented autonomy from Earth's immediate support, and the pressing need to collaborate across diverse, multicultural teams [2]. Indeed, space agencies and commercial enterprises alike have shifted their selection and training processes to cultivate and evaluate a new breed of spacefarer—one who can excel amid uncertainty, lead or follow depending on mission requirements, and pivot seamlessly between the roles of engineer, scientist, caregiver, and communicator.

4.1 Cognitive and psychological resilience

Arguably the most critical attribute of the twenty-first-century astronaut is the ability to adapt cognitively and psychologically to the uncertainties of space travel. Beyond the acute physical demands of launch and microgravity, these explorers face prolonged isolation, disrupted circadian rhythms, and the stress of living in high-stakes environments where errors could prove catastrophic [3]. Under such conditions, even minor interpersonal conflicts or operational challenges can rapidly escalate, making emotional regulation a key determinant of mission success. Astronauts must draw upon a deep reserve of self-awareness and stress-management techniques to identify and mitigate negative emotional states before they compromise judgment or group cohesion [4]. Strategies might include mindfulness, structured social interactions, scheduled downtime for recreation, and even telemedical psychological support to help participants manage stress in real time.

Another facet of cognitive resilience is decision-making under uncertainty. Whether troubleshooting an ailing life-support system or responding to unexpected solar radiation events, astronauts must think critically, weigh incomplete data, and act decisively [1]. Training regimens today often incorporate high-fidelity simulations that deliberately introduce ambiguous or evolving parameters, forcing crews to work under pressure and continually reassess their strategies. Such exercises target cognitive flexibility, or the capacity to switch between tactical concerns (e.g., immediate mechanical failures) and strategic considerations (e.g., the overall mission timeline) with minimal cognitive overload [5]. Effective problem-solving in deep-space or lunar environments requires a keen balance of methodical analysis and inventive improvisation—a mental agility born of rigorous preparation, seasoned with creativity, and grounded by emotional poise.

Equally pivotal is the ability to manage long-duration missions, which test an astronaut's resilience over months or even years in confined habitats [3]. Crew members must develop the psychological stamina to endure sensory monotony—seeing the same walls or planetary surface every day—and the emotional weight of separation from family and friends. Prolonged isolation can give rise to heightened anxiety, depressive symptoms, or interpersonal tensions, making robust support networks and team-building exercises integral to mission design [4]. Training in conflict resolution and active listening helps astronauts address frictions openly and productively, mitigating the risk of interpersonal breakdowns in an environment where cooperation is mandatory for survival. Overall, the capacity to maintain equilibrium under protracted stress is emerging as one of the most sought-after traits in modern astronaut selection.

4.2 Interdisciplinary expertise

In the era of space shuttles and the International Space Station, astronauts began to serve in highly specialized roles—flight engineers, mission specialists, and payload commanders—designed to tackle everything from satellite deployments to biomedical research [6]. Today's spacefarers, however, must integrate multiple disciplines in a single mission profile, blurring the lines between professional categories. With rapid advancements in technology and the multiplication of mission objectives, crew members often function as scientists, engineers, medical officers, and communicators all at once [2]. This interdisciplinary expectation is not merely a matter of broad training; it requires the capacity to think holistically about diverse technical systems and the underlying science, as well as to troubleshoot those systems in real-time emergencies. Consequently, many astronauts hold advanced degrees in fields such as astrophysics, microbiology, robotics, or planetary geology, ensuring the crew's collective knowledge base spans the gamut of potential in-flight tasks.

Another emerging aspect of interdisciplinary expertise is the astronaut's role as a researcher-communicator. Missions now frequently involve experiments designed to advance our understanding of Earth, space environments, or the human body under microgravity conditions. Astronauts must not only execute these experiments with precision but also document their results and interpret findings in collaboration with ground-based scientists [1]. Their ability to share knowledge—both formally in scientific conferences and informally via social media or public outreach—helps extend the impact of each mission, stimulating broader interest in space exploration and inspiring future generations to pursue science, technology, engineering, and mathematics (STEM). Moreover, in deep-space scenarios or on remote planetary outposts, astronauts may act as the first line of research investigators, collecting samples, analyzing data, and making real-time discoveries that cannot wait for Earth-based guidance [3]. Such a role demands a high degree of autonomy and scientific literacy, reinforcing the imperative for well-rounded, interdisciplinary competencies.

Finally, medical and life-support considerations also reinforce the value of cross-training. With limited or delayed access to Earth-based medical facilities, astronauts need sufficient medical training to handle emergencies, perform routine health check-ups, and assist with psychological support [5]. This requirement applies as much to short-duration commercial flights—where an onboard medical contingency might arise—as it does to multi-year missions across interplanetary space. Since crew survival depends on the functionality of life-support systems and the stability of individual well-being, astronauts are increasingly expected to comprehend and, if necessary, repair essential habitat elements, from water recycling units to oxygen generators. In a realm where no single specialist may be readily available, each astronaut becomes a multifunctional asset, and collaborative knowledge-sharing forms the backbone of safe and efficient operations.

The twenty-first-century astronaut thrives on a foundation of cognitive resilience and expansive interdisciplinary expertise, equipped not only to pilot advanced spacecraft but also to endure psychological strains and bridge multiple scientific domains [2, 6]. This evolution reflects the changing face of space travel, where missions no longer revolve around short flights or simple orbital tours but address pressing research questions and pave the way for permanent off-world settlements. As technology, commerce, and strategic priorities continue to reshape the mission architecture, the skill profile of the astronaut will remain in constant flux—requiring ever greater adaptability, creativity, and collaboration. Future decades may see these explorers working alongside artificial intelligences, mining lunar regolith for resources, or cultivating plants in Martian greenhouses, all within the same expedition. Consequently, the ability to navigate and synthesize knowledge across disciplines will be fundamental, ensuring that humans continue to play an irreplaceable role in charting our collective future in space.

5. The psychology of space travel: Cognitive and emotional demands across different astronaut profiles

The psychological landscape of space travel varies significantly depending on the individual's objectives, mission duration, and the broader context of their journey. While all astronauts grapple with the challenges of microgravity, isolation, and life-threatening environments, the nature of these stressors—and how they manifest in daily activities—differs among career astronauts, short-term visitors, and prospective settlers [1]. Each profile demands specific forms of mental preparation, emotional regulation, and interpersonal adaptability, reflecting the spectrum of experiences that exist within modern space exploration. From highly trained professionals handling scientific research on long-duration missions to space tourists seeking a once-in-a-lifetime adventure, every traveler must develop robust coping strategies to navigate disorientation, fear, and the pressures of high-stakes environments [2]. By examining these distinct profiles, we gain insight into the varied psychological demands of human spaceflight and the nuanced approaches required for success in each role.

5.1 The professional astronaut

Professional astronauts—often employed by space agencies or commercial entities—occupy the most traditional category in this evolving landscape. Their extensive training covers advanced vehicle operations, scientific research protocols, and emergency preparedness, requiring them to function as both specialists and generalists [3]. In this context, cognitive load management becomes paramount, as these individuals must juggle complex mission objectives, troubleshoot intricate systems, and respond effectively to dynamic situations such as docking maneuvers or extravehicular activities (EVAs). The psychological pressure is intensified by the knowledge that any miscalculation could jeopardize not just the mission but also the lives of fellow crew members, prompting astronauts to refine stress-management techniques and reinforce collective trust [1]. Team cohesion is further challenged by diverse cultural backgrounds; multinational crews must navigate linguistic differences, cultural nuances, and varying interpersonal norms to maintain operational harmony [4].

Professional astronauts typically undergo rigorous, multi-year training programs following selection. For instance, NASA and ESA candidates often complete two years of basic training before being eligible for assignment, with additional mission-specific preparation that can extend for several more years depending on flight schedules and mission complexity [4]. Notably, some ESA astronauts have flown many years after their initial selection, sometimes with long gaps between training phases. Training includes advanced technical instruction, spacecraft systems, extravehicular activity (EVA) simulations, scientific experiment protocols, and psychological resilience conditioning. Despite this intensity, flexibility in timing and readiness remains essential, as mission needs and international coordination frequently affect crew assignment timelines.

Moreover, professional astronauts endure the harsh realities of isolated living conditions, whether aboard the International Space Station (ISS) or within future deep-space habitats. Prolonged exposure to microgravity disrupts bodily functions and can contribute to fatigue, which, if unaddressed, may impair both cognitive and emotional well-being [3]. Social interactions occur in a small, enclosed environment, leaving little room for privacy or escape from interpersonal tensions—factors that can erode morale over time if not managed proactively. Consequently, agencies integrate psychological resilience into selection criteria and training regimens, employing simulations and stress tests designed to evaluate how candidates respond under prolonged duress [1]. Frequent communication with ground support, routine psychological check-ins, and structured recreation play pivotal roles in helping these astronauts maintain balanced mental health. By excelling at complex problem-solving while preserving interpersonal harmony, professional astronauts exemplify the multifaceted responsibilities of twenty-first-century space exploration.

5.2 The space tourist

Space tourism, once the domain of speculative fiction, has emerged as a tangible reality through suborbital flights and orbital sojourns facilitated by commercial companies [2]. Space tourists typically come from varied professional backgrounds—business, entertainment, sports, or academia—and may have limited experience with rigorous aerospace or scientific endeavors. Their psychological preparation thus focuses on the disorienting and highly emotional aspects of brief space travel: from rapid acceleration during launch to the visceral awe of viewing Earth from orbit. Short, intensive training regimens must equip them to handle potential anxiety, motion sickness, or unforeseen technical glitches without succumbing to panic [5]. In many cases, space tourists also confront heightened public attention, adding performance pressure and media obligations to their mental burden.

Sensory overload is a frequent concern for first-time travelers, as the transition from terrestrial life to microgravity can induce feelings of euphoria, disorientation, or even fear. Astronaut trainers emphasize coping strategies that help participants modulate their emotional responses, such as controlled breathing exercises or structured focus on routine tasks [2]. At the same time, space tourists must navigate expectations shaped by popular culture, which may not always align with the realities of cramped spacecraft and strict safety protocols. Reintegrating into normal life post-flight poses another psychological challenge, as the awe and excitement of space travel can leave individuals longing for the exceptional experience they have just left behind [1]. While their missions are relatively short compared to professional astronauts, space tourists still encounter intensive emotional demands that call for adaptability, self-awareness, and an openness to learning from seasoned experts.

5.3 The one-way colonist

Perhaps the most extreme astronaut profile is that of the one-way colonist—a pioneer ready to establish a permanent presence on a distant planetary body such as the Moon or Mars. Unlike professional astronauts or space tourists, these individuals may face the existential weight of potentially never returning to Earth, which imposes profound psychological ramifications [3]. The emotional challenges include leaving behind family, friends, and the familiar comforts of terrestrial life, as well as confronting a future shaped by extreme isolation and sparse social networks. In such a scenario, a colonist's mental stamina must endure not merely months but possibly years or a lifetime in an alien environment. The deep sense of purpose that often accompanies off-world settlement helps sustain motivation, yet long-term contentment also depends on cultivating community bonds, meaningful work roles, and reliable channels for communication with Earth [1].

The logistical demands of one-way missions amplify stress in ways unimaginable in shorter excursions. Colonists must learn to manage habitat maintenance, resource scarcity, and emergency protocols without immediate ground support, making resiliency and adaptability indispensable traits [4]. The challenges of partial gravity, hazardous radiation levels, and the psychological toll of a monotonous environment intensify the need for robust mental health support systems. Crew autonomy drives the development of advanced medical, psychological, and social coping strategies, with each member expected to contribute to communal well-being as well as mission objectives [2]. As these inhabitants become de facto stewards of humanity's extraterrestrial expansion, their ability to forge a sustainable culture in a remote, unforgiving setting will significantly influence the success or failure of permanent settlements beyond Earth's orbit. Consequently, the one-way colonist's psychological journey represents one of the boldest frontiers of human exploration, demanding a combination of grit, adaptability, and a well-grounded sense of personal meaning [3].

Table 1 illustrates key psychological demands of space travel based on the traveler's role and the nature of their mission; it is not intended to be exhaustive.

Category	Professional astronaut	Space tourist	One-way colonist
Typical mission duration	Several days to several months (e.g., ISS missions) [5], with some stays up to a year or more [3].	Usually a few days or less, typically involving suborbital or short orbital flights [6].	Multi-year or permanent missions with no guaranteed return to Earth [3].
Key stressors	• Prolonged confinement and isolation [1, 2] • High operational demands [3] • Interpersonal friction in small crews [7]	• Rapid environmental changes [2] • Anxiety over unfamiliar sensations, performance pressure [6] • Possible "sensory overload" [5]	• Extreme isolation and great distance from Earth [1, 3] • Psychological burden of indefinite separation [7] • Dependence on limited resources for survival [8]
Training intensity	Rigorous, multi-year programs [5] emphasizing vehicle operations, scientific tasks, emergency preparedness, and group cohesion [1].	Condensed training oriented toward safety, zero-g adaptation, essential emergency protocols [4].	Intensive and specialized to handle permanent settlement tasks, life-support maintenance, and psychological endurance [3, 9].
Coping strategies	• Structured counseling, scheduled rest cycles [2] • Conflict-resolution training, mindfulness [1] • Ongoing support from mission control [8]	• Short-term orientation to manage acute stressors [6] • Controlled breathing, basic emotional regulation [1] • Guidance from professional astronauts [4]	• Emphasis on autonomy and community-building [7] • Continuous mental health support, including onsite counseling [8] • Deep sense of purpose and commitment [3]
Primary objectives	• Scientific research and experimentation [3] • Station or vehicle maintenance [5] • Operational readiness and crew safety [1]	• Personal fulfillment, brand ambassadorship [6] • Public engagement or philanthropic motives [2] • Rapid experiential missions [5]	• Laying groundwork for permanent off-world communities [3] • Resource extraction, habitat construction [6] • Long-term societal and scientific development [7, 8]

Category	Professional astronaut	Space tourist	One-way colonist
Psychological preparation	• Advanced simulations (e.g., analog missions) [1] • Comprehensive psychometric assessments [2] • Team-building and cultural awareness training [7]	• Basic psychological screening for acute issues [4] • Brief sessions on emotional regulation, motion sickness prevention [2]	• In-depth screening for resilience and adaptability [3] • Ongoing psychological support for indefinite durations [8]
Reintegration challenges	• Readjustment to Earth gravity [5] • "Downlift" after intense focus [1] • Reconnecting with family and normal life routines [2, 7]	• "Come-down" from a once-in-a-lifetime experience [6] • Possible mismatch of expectations vs. reality [4] • Limited official support post-flight [5]	• Often little or no return to Earth, requiring perpetual adaptation [3] • Existential, social, and cultural implications of forming new societies [1] • Sustained psychological strain [7]
Training duration & pathways	• Multi-year selection and training programs; often includes 2 years of basic training followed by mission-specific instruction, with some astronauts flying many years post-selection (e.g., ESA) [5]	• Condensed training over a few days to weeks; focuses on safety, microgravity adaptation, and emergency basics [4]	• Specialized, long-term preparation for habitat maintenance, autonomy, psychological endurance, and group dynamics [3]

Table 1.
Comparative overview of astronaut profiles and psychological needs.

6. Psychological resilience and adaptability: Core traits of the future astronaut

As human space exploration expands beyond low Earth orbit and ventures toward the Moon, Mars, and potentially even deeper into the solar system, the importance of robust psychological resilience and adaptability has never been greater [1]. In space, where immediate evacuation or rescue may not be possible, even minor emotional or cognitive disruptions can escalate into serious risks for mission success and crew well-being. Consequently, astronaut training programs and selection criteria now emphasize the mental and interpersonal qualities required to handle stress, navigate complex emergencies, and maintain effective collaboration over extended durations [2]. These qualities manifest in several core competencies: the capacity to regulate emotions under duress, the agility to solve problems in unpredictable circumstances, the skill to foster positive social dynamics in close quarters, and the sense of purpose that sustains motivation through lengthy or even indefinite missions [3]. Together, these traits form the psychological bedrock upon which future astronauts will rely as they pioneer the next era of human spaceflight.

6.1 Emotional regulation and stress management

One of the most crucial attributes for an astronaut in any mission scenario is the ability to manage and regulate emotional responses in the face of unrelenting stress. Launch and re-entry phases alone subject crew members to intense physical forces and heightened risk, requiring careful composure to avoid panic or disorientation [1]. However, stressors do not end once a spacecraft achieves orbit; prolonged missions introduce cycles of uncertainty, routine disruptions, and the mental fatigue associated with near-constant vigilance. Psychological support systems—ranging from mission control check-ins to telemedical consultations—assist astronauts in coping with high-pressure situations, yet much responsibility still lies with individual crew members to recognize and quell rising anxiety or frustration [2]. Techniques such as mindfulness, controlled breathing, and emotional de-escalation have gained traction in modern astronaut training, empowering individuals to reset their mental states before stress accumulates to detrimental levels [4]. By maintaining emotional equilibrium, astronauts not only ensure clear thinking during emergencies but also uphold a stable social environment within the spacecraft or habitat.

6.2 Cognitive flexibility and problem-solving in space

The capacity for rapid, effective problem-solving ranks among the most valued competencies for crews operating in harsh, isolated conditions. Mission demands can shift unexpectedly as technical glitches, resource shortages, or environmental hazards emerge—factors that become more acute the further a crew travels from Earth's ready support [3]. Cognitive flexibility enables astronauts to reframe tasks, devise contingency plans, and integrate real-time data, often under strict time constraints. High-fidelity simulations and frequent drills cultivate this adaptability, requiring trainees to make swift decisions with incomplete or shifting information [5]. In conjunction with advanced automation, astronauts also refine their judgment on when to trust artificial intelligence inputs and when to rely on human intuition, adding another layer of complexity to problem-solving processes [1]. By mastering cognitive flexibility, future astronauts can pivot seamlessly from routine tasks to emergency response, ensuring the crew's collective survival and the mission's ultimate success.

6.3 Social and interpersonal dynamics

No matter how advanced spacecraft or habitats become, the intimate human interactions within them remain a pivotal element of mission stability and morale. Confinement in restricted spaces for weeks, months, or even years intensifies everyday interpersonal dynamics—whether it be cultural differences, habit conflicts, or personality clashes [2]. Consequently, astronauts require refined communication skills, empathy, and conflict resolution techniques to sustain functional relationships under these compressed conditions. Space agencies now emphasize training in collaborative problem-solving and structured social activities, recognizing that unresolved tensions can corrode unity and diminish overall performance [4]. In particular, multinational missions demand cross-cultural competence, as crew members must navigate language barriers and distinct social norms while working toward shared objectives [5]. Effective interpersonal dynamics not only mitigate stress but also create an environment in which innovation and mutual support can flourish, both vital components for exploratory missions in remote, unforgiving regions of space.

6.4 Sense of purpose and psychological well-being

At the heart of long-duration or permanent space missions lies a profound psychological requirement: the need for an enduring sense of purpose. Extended travel to Mars or the establishment of lunar outposts can involve years of relative isolation, repetitive tasks, and limited sensory stimulation [3]. Without a deep personal commitment to the mission's broader scientific, exploratory, or societal goals, crew members risk succumbing to apathy, low morale, or even depression. Accordingly, agencies like NASA now incorporate discussions of mission meaning, individual aspirations, and community-building into both selection and training processes [1, 4]. Anchoring each crew member's experience to a greater cause—be it advancing human knowledge, inspiring future generations, or laying the groundwork for permanent off-world settlements—can buffer against psychological fatigue. Where purpose flourishes, astronauts frequently exhibit high levels of resilience, harnessing motivation to surmount the daily rigors of space life and maintain a hopeful outlook, even under daunting circumstances [6].

7. Training for psychological preparedness: Building the future astronaut's mindset

Achieving psychological readiness for space travel is a complex process that intertwines rigorous mental conditioning with continual self-reflection and adaptation. Modern astronaut training goes beyond mastering spacecraft systems or physical endurance; it emphasizes developing robust coping mechanisms for isolation, stress, and unexpected mission complications [1]. Whether preparing career astronauts for extended assignments on the International Space Station, commercial crew members for shorter orbital flights, or prospective settlers bound for the lunar surface or Mars, agencies and companies alike understand that psychological health is paramount to mission success. As a result, training strategies incorporate realistic simulations, stress inoculation exercises, and cognitive-behavioral techniques aimed at fortifying resilience under pressure. By systematically building these mental defenses and fostering supportive team cultures, space programs help ensure that crews can handle the emotional, interpersonal, and operational challenges inherent to venturing beyond Earth.

7.1 Psychological selection criteria for different astronaut profiles

Prior to embarking on any formal training program, space agencies and commercial organizations employ rigorous selection processes to filter candidates based on their psychological suitability. For professional astronauts—tasked with scientific research, complex operational duties, and potential long-duration flights—examiners often use psychometric evaluations, group exercises, and structured interviews to assess stress tolerance, emotional regulation, and team compatibility [2]. In contrast, space tourists or private individuals with more limited mission objectives may undergo abbreviated psychological screenings, ensuring they can cope with the sensory overload and potential anxiety of microgravity without the comprehensive skill demands of a career astronaut [3]. Prospective settlers, who face the unique likelihood of long-term or permanent habitation beyond Earth, undergo even more intensive evaluations that probe their motivations, sense of purpose, and capacity for enduring monotony or social isolation [1]. Across these divergent profiles, the unifying priority remains identifying individuals with the mental fortitude and interpersonal agility essential for survival and collective harmony in extraterrestrial environments.

7.2 Training the mind for space

Training for psychological resilience in space frequently employs analog missions, virtual reality, and AI-driven simulations designed to mimic the stresses and constraints of remote habitats. Simulated long-duration missions—whether in terrestrial facilities like the Hawai'i Space Exploration Analog and Simulation (HI-SEAS) or polar research stations—offer crews an immersive preview of what it means to live, work, and problem-solve in tight quarters [4]. The aim is to refine coping strategies under realistic conditions, forcing participants to manage conflict, divide limited resources, and maintain productivity with minimal external support. Alongside these immersive experiences, cognitive behavioral techniques and mindfulness practices help trainees build awareness of their emotional triggers and develop healthy responses to stress [5]. By pairing practical simulations with psychological frameworks, astronaut training programs cultivate a mindset adaptable to the inherently unpredictable nature of spaceflight and extended off-world living.

7.3 Enhancing team cohesion and conflict resolution skills

Finally, preparing astronauts for the interpersonal dimension of spaceflight is just as critical as honing technical aptitudes or stress-management strategies. Crews confined to a capsule or habitat may find small frictions magnified, as limited personal space and constant togetherness can erode group morale [6]. To address these realities, training regimes increasingly focus on collaborative problem-solving, active listening drills, and role-playing scenarios that place trainees in tense interpersonal situations. By practicing constructive communication and structured de-escalation techniques, participants learn to identify the root causes of conflict and reconcile differences swiftly for the sake of mission continuity [1]. These skills are particularly vital in long-duration or multicultural crews, where varying cultural norms and communication styles can lead to misunderstandings. When astronauts develop the competencies to navigate conflict compassionately and efficiently, they not only safeguard mental health but also optimize overall mission performance in the unforgiving environment of space.

8. Long-term mental health and psychological risks in space

Long-duration missions beyond Earth's orbit pose profound challenges to astronaut well-being, with psychological stressors equal in magnitude to the physical rigors of space travel [1]. Unlike short orbital stays, multi-month or multi-year expeditions to Mars or other distant destinations subject crews to sustained isolation, sensory monotony, and a heightened awareness that rescue or return options may be significantly delayed [2]. These circumstances amplify the importance of proactive mental health strategies, as unmitigated stress or interpersonal conflict can derail mission objectives and jeopardize individual safety. Space agencies have therefore ramped up research and training efforts to address these risks—bridging psychology, neuroscience, and engineering solutions to support astronaut resilience [3, 4]. At the same time, commercial and international collaborations broaden the scope of participants in deep-space missions, emphasizing the need for culturally inclusive approaches to managing human factors in cramped, isolated habitats [5].

8.1 The impact of isolation and confined environments

Extended periods in closed habitats—be they aboard spacecraft, orbiting stations, or planetary outposts—can induce substantial psychological strain. With minimal privacy and limited environmental variety, astronauts risk developing irritability, anxiety, or depressive symptoms [6]. The forced proximity magnifies mundane tensions among crewmates, compelling them to confront every disagreement within a confined area, often under high operational pressure. Research from simulated missions in environments like the Antarctic and analog stations such as HI-SEAS consistently demonstrates that strong group cohesion and adaptive coping strategies are critical for sustaining morale [2, 7]. Crew members who communicate openly, share responsibilities fairly, and schedule purposeful recreational activities—such as collective meals or guided mindfulness sessions—tend to navigate the stress of isolation more effectively than those who lack defined social support structures [1, 7].

NASA's Behavioral Health & Performance Program (BHP) underscores that proactive mental health measures must be integrated into mission planning, including support from ground-based psychologists, structured rest periods, and crisis response protocols [8]. Effective design of living quarters can also alleviate some psychosocial issues, as providing distinct "quiet spaces" and personal stowage areas reduces tension in a microgravity or reduced-gravity habitat. By refining both architectural and procedural elements, mission planners seek to optimize group function, anticipating that prolonged isolation in deep-space operations will only heighten the need for robust, integrated strategies to safeguard mental health [4].

8.2 Neurocognitive changes in microgravity and space travel

Microgravity exerts notable effects on neurological and cognitive functioning, from shifts in spatial orientation to alterations in sleep patterns [1]. Astronauts frequently experience disrupted circadian rhythms, a phenomenon intensified by frequent day-night cycles in low Earth orbit or the absence of natural light cues during interplanetary travel [2]. Sleep deprivation and fragmented rest can compound stress, eroding attention span, memory, and emotional regulation over time. Studies involving both real and simulated missions suggest that these neurocognitive changes may become more pronounced the longer an individual remains off-world, potentially complicating routine tasks and emergency responses [1, 9].

Countermeasures range from carefully calibrated lighting systems—designed to mimic natural daylight cycles—to enforced sleep hygiene protocols that prioritize consistent rest periods [8]. Exercise routines and dietary management further support cognitive function by mitigating some physiological stressors of microgravity, such as bone density loss and muscle atrophy. Ongoing research also investigates pharmaceutical or nutritional interventions that might sustain cognitive performance on interplanetary missions, though their long-term efficacy remains under study [9]. Nevertheless, robust crew collaboration, frequent self-assessment, and a high degree of situational awareness can all help astronauts adapt to neurocognitive fluctuations, ensuring mission-critical activities proceed safely and efficiently.

While all astronauts face a range of psychological risks in space, the severity and likelihood of these risks vary significantly by mission type, duration, and crew composition. NASA's *Human Research Roadmap* categorizes these hazards according to their potential impact on health and performance across different mission scenarios (e.g., low-Earth orbit vs. planetary missions). This framework recognizes that short-duration space tourism may carry lower psychological risk than extended exploration missions, which amplify stressors like isolation, sensory monotony, and communication delays [10].

As well as summarizing major risks, psychological manifestations, and potential mitigation strategies, **Table 2** integrates this risk-ranking perspective by mapping

Risk factor	Potential psychological manifestations	Potential mitigation strategies	Severity likelihood	Most relevant astronaut profile(s)
Isolation & confinement	• Elevated stress, anxiety, irritability [1, 2] • Depression and withdrawal due to social monotony [8] • Heightened conflict in tight quarters [7]	• Structured daily schedules, communal activities [3] • Ground-based psychological support (telemedicine) [8] • Transparent conflict-resolution protocols [7]	High High [10]	All, especially Professional & Colonist
Sensory monotony	• Cognitive fatigue, reduced attention span [1] • Boredom and diminished motivation [5]	• Rotating task assignments for variety [1, 3] • Use of virtual or augmented reality to provide novel "escapes" [9] • Group leisure and shared hobbies [7]	Moderate–High High [10]	Colonist, Professional
Radiation exposure	• Heightened fear or anxiety about long-term health effects [6, 9] • Chronic stress related to unknown risks [2]	• Adequate shielding and protective equipment [6] • Informed consent and clear communication of hazards [1] • Psychological counseling for stress management [8]	High Medium (varies by mission) [10]	Colonist, Professional (deep space)
Microgravity & physiological changes	• Sleep disruptions, circadian rhythm misalignment [2, 4] • Mood swings or irritability due to fatigue [7]	• Enforced sleep–wake cycles with specialized lighting [8] • Regular exercise protocols (treadmill, resistance) [3] • Pharmaceutical and nutritional support [5]	Moderate High [10]	All
Extended communication delays	• Frustration or stress from lag in conversation with Earth [1] • Feelings of isolation, being "cut off" from real-time feedback [7]	• Pre-scripted protocols for emergency comms [6] • Autonomous decision-making training [9] • Regular asynchronous video messages to maintain social contact [8]	High (for deep-space) Medium [10]	Colonist, Long-duration Professional

Table 2.
Major psychological risks in long-duration space missions.

specific risks to astronaut profiles and estimating their severity and likelihood, based on current research and operational experience. It is not intended to be exhaustive.

8.3 Psychological and ethical challenges of permanent space settlement

Future endeavors to establish permanent habitats on the Moon or Mars introduce additional layers of psychological and ethical complexity. Individuals who embark on one-way or semi-permanent assignments face the reality of leaving Earth indefinitely, bringing existential dilemmas and an intensified need for emotional support [3]. Loneliness, grief over missed life events, and the potential for feeling "forgotten" by Earth-based society become more acute in these scenarios. Ethical questions also arise regarding informed consent, particularly if emerging health data suggests that deep-space radiation or protracted microgravity exposure may have unknown long-term consequences for mental health [1, 2].

In constructing a sustainable off-world community, crew members transition from mere travelers to pioneers, shaping social norms and cultural practices in an alien context [6]. Designing mechanisms for conflict resolution, governance, and mental health care thus becomes vital for cohesive community-building [5]. The psychological demands of permanent settlement demand ongoing research, including longitudinal studies of individuals living in isolated or extreme environments on Earth. Continuous monitoring and prompt intervention by both in-situ medical personnel and Earth-based mental health professionals will be crucial for preventing the kind of chronic stress or depression that could unravel a settlement's viability [4, 8]. Ultimately, the goal is to equip settlers not only with the technical means to survive in unforgiving conditions but also the robust psychological frameworks necessary for a meaningful, stable life beyond Earth.

9. Challenges and ethical considerations

The expansion of human spaceflight to commercial enterprises, deep-space missions, and permanent off-world settlements presents a host of moral, legal, and practical dilemmas. No longer confined to the relatively narrow scope of government-sponsored, short-duration orbital campaigns, space exploration now involves private tourism, multinational collaborations, and speculative ventures for colonization on the Moon or Mars [1, 2]. These evolutions raise fundamental questions about how to protect crew health in the long term, manage interactions between humans and advanced artificial intelligence, draw upon lessons from extreme terrestrial environments, and ensure equitable access to the final frontier. Viewed through a broad ethical lens, space exploration is not merely a technical challenge but also a profound social undertaking, bringing human values and vulnerabilities into stark focus. In this realm, responsible policies and practices will be critical for stewarding an era in which humanity transcends its planetary origins while preserving fundamental principles of safety, justice, and cooperation [3, 4].

9.1 Health risks and ethical concerns

One of the most pressing concerns in space exploration is the long-term health risk associated with radiation exposure, microgravity, and confinement. Data gathered from International Space Station (ISS) expeditions show that bone density loss, cardiovascular strain, and immune system dysregulation pose persistent threats to astronaut well-being [5]. Extended missions to the Moon and Mars amplify these risks, as crews lack rapid evacuation options and must cope with higher radiation levels, partial gravity, and limited medical facilities. Proposals for permanent settlements further highlight ethical dilemmas around informed consent: individuals agreeing to one-way missions or multi-year residencies may face medical outcomes that are impossible to predict based on current research [6]. Additionally, the genetic effects of cosmic radiation and the psychological toll of protracted isolation spark debates about whether humans have a moral obligation to limit such risks or whether the spirit of exploration justifies pushing these boundaries. As private entities take on a larger role in mission planning, regulatory oversight becomes essential to ensure transparency about health hazards and the protective measures provided to all crew members [1].

The need for adequate mental health care reinforces these ethical imperatives. Chronic stress, sleep disturbances, and interpersonal tension can severely compromise astronauts' decision-making and group cohesion, placing entire crews at risk [7, 8]. Agencies such as NASA's Human Research Program have prioritized behavioral health support in their mission frameworks, yet the challenge grows more acute in the isolated domains of cislunar or interplanetary space [9]. Questions arise about who bears responsibility for psychological care in private missions and whether commercial pressure might lead to underestimating the emotional resources required for safe operation. Striking a balance between bold exploration and respecting the limits of human physiology and mental well-being remains a central ethical tension, underscoring the need for stringent protocols and continuous medical surveillance throughout long-duration flights.

9.2 The balance between human and AI contributions

Advances in artificial intelligence and robotics stand poised to reshape the astronaut's role, raising ethical questions about where and how human expertise should remain indispensable. Highly automated systems already assist with navigation, life-support diagnostics, and onboard data analysis, reducing crew workload and mitigating certain human error factors [1, 11]. However, overreliance on AI may engender new vulnerabilities, especially if an unforeseen software glitch or cyberattack cripples critical subsystems far from Earth-based repair resources. This scenario forces mission planners to weigh the benefits of efficiency against the imperative of human oversight, ensuring that astronauts retain the skills and authority to override automated decisions when necessary [2, 11].

On deeper philosophical levels, the integration of AI touches upon concerns about devaluing the very essence of human exploration. Historically, spaceflight has symbolized human courage, ingenuity, and the unyielding quest to push personal limits in challenging realms [12]. Substituting machines for human presence might undermine those values, reducing space travel to a purely operational or commercial enterprise lacking the inspirational power that comes from humans venturing forth themselves. Thus, spacefaring societies face a moral balancing act: leveraging AI and robotics to optimize safety and performance while preserving the unique cognitive and emotional contributions only people can offer. In future missions, ethical codes and mission designs must safeguard human autonomy, clarifying lines of authority between crew members and AI systems—even as the synergy of both will undoubtedly shape successful ventures in deep space.

9.3 Lessons from terrestrial extreme environments

Polar expeditions, deep-sea research habitats, and remote military deployments provide valuable analogs for understanding the psychological and logistical demands of extended missions beyond Earth [7, 13]. Individuals operating in these extreme terrestrial settings often share challenges akin to those of astronauts—chronic isolation, resource scarcity, hazardous conditions, and the need for tightly coordinated teamwork. Strategies that have proven effective in these environments include well-structured leadership models, peer support networks, and robust emergency preparedness [8]. Such approaches underscore the importance of deliberate community-building measures, periodic mental health check-ins, and protocols for conflict resolution and crisis management.

Yet, direct parallels can be misleading: even the most remote terrestrial environments still offer the possibility of rescue or eventual return. The distance factor in deep-space travel amplifies the psychological and ethical stakes. Nonetheless, integrating best practices from polar or undersea missions can inform the design of space habitats, training regimens, and in-flight support structures [13]. These lessons, scaled up for cosmic distances, emphasize the human capacity for resilience and adaptation when provided with supportive frameworks. Drawing on terrestrial precedents, mission planners can craft robust strategies for survival and well-being, refining them to meet the unique perils and scale of interplanetary travel.

9.4 Sociopolitical implications of space colonization

As space exploration transitions toward settlement, its sociopolitical dimensions become increasingly prominent. Questions about who "gets to be an astronaut" and how nations or corporations allocate resources for off-world development intersect with broader issues of equity, representation, and potential exploitation [2]. While open calls for civilian astronauts have democratized access somewhat, the financial barriers to space tourism and the technological advantages certain nations hold can exacerbate global inequalities, effectively limiting off-world opportunities to elite participants [14]. These disparities risk reproducing Earth-bound power imbalances on the lunar or Martian frontiers, sparking debates on how to govern new extraterrestrial communities responsibly.

International treaties like the Outer Space Treaty of 1967 articulate broad ideals—declaring space the "province of all mankind"—yet the practical governance of space resources, property rights, and workforce ethics remain unresolved [15]. Settlement scenarios, particularly those involving private or joint public-private endeavors, might encourage competition over scarce materials or prime land for bases. If unregulated, such competition could lead to conflicts mirroring territorial disputes on Earth, undermining the collaborative ethos that initially spurred human space exploration. Consequently, policymakers and mission architects bear an ethical duty to craft frameworks that foster cooperation, ensure equitable access, and preserve scientific transparency. Such deliberate approaches will be essential for preserving both the safety and integrity of emerging space societies in the decades to come.

10. Conclusion: The mind as the ultimate spacefaring tool

As humanity stands on the threshold of unprecedented exploration—encompassing commercial flights, interplanetary voyages, and permanent extraterrestrial

DOI: http://dx.doi.org/10.5772/intechopen.1010596

habitats—the psychological dimension of space travel emerges as both the unifying thread and the ultimate frontier [1]. Technical achievements such as reusable rockets, advanced life-support systems, and automated navigation have brought interplanetary targets within closer reach, yet the demands on human cognition, emotion, and social interaction remain intensely complex [2]. The willingness to confront unknown dangers, manage long-term isolation, and work collaboratively with both human and artificial intelligence agents underscores how mental adaptability can determine the success or failure of even the most advanced missions [3]. In this evolving context, "the Right Stuff" transcends the rugged heroism of the past: it is now a multifaceted construct comprising resilience, teamwork, empathy, and a persistent sense of purpose. These attributes allow astronauts—and all individuals venturing into space—to thrive in climates of adversity, protecting not only mission objectives but also the fundamental well-being of every crew member.

Ultimately, the psychological readiness to adapt, innovate, and connect with others in confined, high-stakes environments stands at the heart of future space exploration. Whether the role involves piloting a commercial craft to low Earth orbit, training as a one-way colonist, or guiding short-term tourists through the wonders of suborbital flight, a well-honed mental toolkit is indispensable [4]. Agencies, research institutions, and private companies now recognize that astronaut training must incorporate extensive mental conditioning, holistic stress management, and advanced interpersonal skills [5, 6]. The collective journey off-world will hinge on fostering not only engineering excellence but also the emotional fortitude to cope with solitude, risk, and perpetual change. Seen in this light, the mind truly becomes the most vital spacefaring instrument, enabling humans to persevere across millions of miles of emptiness and forge thriving outposts amid the stark beauty of distant worlds [7, 8].

Acknowledgements

The Open Access Publishing Fee for this chapter is generously funded by the College of Arts & Sciences, Embry-Riddle Aeronautical University – Worldwide.

Note

In this chapter, the term *astronaut* is used inclusively to refer to all individuals who travel into space, regardless of national or cultural designation. This includes *cosmonauts* (Russia), *taikonauts* (China), *spationauts* (France), *angkasawan* (Malaysia), *vyomanauts* (India, proposed), and *afronauts* (used informally in African space contexts).

Author details

Donna L. Roberts
Embry-Riddle Aeronautical University – Worldwide, Daytona Beach, FL, USA

*Address all correspondence to: donna.roberts@erau.edu

References

[126] Kanas N, Manzey D. Space Psychology and Psychiatry. 2nd ed. Dordrecht: Springer; 2008

[127] Santy PA. Choosing the Right Stuff: The Psychological Selection of Astronauts and Cosmonauts. Santa Barbara, CA: Praeger; 1994

[128] NASA. NASA's Journey to Mars: Pioneering Next Steps in Space Exploration. Washington, D.C.: NASA Headquarters; 2015

[129] NASA. Astronaut Requirements. 2017. Available from: https://www.nasa.gov/humans-in-space/astronauts/astronaut-requirements/

[130] NASA. Astronaut Fact Book. 2022. Available from: https://www.nasa.gov/wp-content/uploads/2015/05/astronaut_fact_book_june_2022.pdf

[131] NASA. NASA's Artemis Program: Humanity's Return to the Moon. 2021. Available from: https://www.nasa.gov/specials/artemis/

[132] Landon LB, Slack KJ, Barrett JD. Teamwork and collaboration in long-duration space missions: Going to extremes. American Psychologist. 2018;**73**(4):563

[133] NASA's Human Research Program: Behavioral Health & Performance. 2025. Available from: https://www.nasa.gov/hrp/human-factors-and-behavioral-performance/

[134] National Academies of Sciences, Engineering, and Medicine. Space Studies Board Annual Report 2019. Washington, D.C.: The National Academies Press; 2021

[135] NASA Human Research Program. Human Research Roadmap: Risks and Gaps. 2024. Available from: https://humanresearchroadmap.nasa.gov

[136] Hambuchen K, Marquez J, Fong T. A review of NASA human-robot interaction in space. Current Robotics Reports. 2021;**2**(3):265-272

[137] Whitman Cobb WN. Public opinion of commercial space exploration. Astropolitics. 2023;**21**(2-3):164-178

[138] Palinkas LA, Suedfeld P. Psychological effects of polar expeditions. Lancet. 2008;**371**(9607):153-163

[139] Pattanayak T. Surveying the landscape for promoting safety, transparency, accountability, and equitable access in commercial space tourism. New Space. 2024;**12**(3):214-222

[140] United Nations Office for Outer Space Affairs. Treaty on Principles Governing the Activities of States in the Exploration and Use of Outer Space, including the Moon and Other Celestial Bodies. 1967. Available from: https://www.unoosa.org/oosa/en/ourwork/spacelaw/treaties.html

Chapter 5

Perspective Chapter: Diversity and Inclusion in Space – The Case for Parastronauts

Eleonore Poli

Abstract

The present and future of space exploration presage an exciting era for astronaut recruitment. With the first parastronaut recruited in 2021 by the European Space Agency, the spacecrafts, space station, and protocols are slowly shifting to more inclusive space missions, paving the way for other parastronauts. On Earth, analogue parastronauts are already breaking barriers by enabling researchers to adapt analogue space mission habitats to fit their disability. While the protocols and modifications are not always transferable to space, the psychological outcomes and crew dynamics provide one-of-a-kind insight into teamwork in extreme environments and situations between astronauts and parastronauts, which not only is crucial for future parastronauts but also could potentially hurt astronauts in long-duration missions. This chapter covers the current ongoing research efforts to make space parastronaut friendly, with its advantages and ramifications, and the crucial role it will play on Earth and the future of space.

Keywords: parastronauts, Mars mission, astronaut selection, inclusivity, space exploration

1. Introduction

There has been perhaps a higher diversity in the types of animals brought to space than there has been in humans. The characteristics of astronauts originally served two purposes: that of accomplishing the mission in space safely and accurately and that of sending a message. Sending the first man in space and the first woman in space was a relevant description of the soviet idealism, compared to the American idealism of sending men, selecting women for the Apollo mission only to discard them to serve an agenda that was critical at the time of the cold war.

Space agencies, such as NASA, which depended on governmental funding and public support, were feeling the pressure as the Apollo missions continued, with outrage at Alan Shepard playing golf on the Moon in the middle of vast Vietnam war spending. Indeed, astronauts are the tip of the iceberg of the vast enterprise of space exploration, the public figure, and as such are the image of space mission, needing to mirror an image of perfect nationalism. This has added a requirement to the many qualities and skills required of astronauts to conduct their mission, the same way Yuri Gagarin was chosen over German Titov to be the first man in space, to reflect proletarian excellence in the time of the Soviet Union.

As missions evolved to requiring test pilots to man new spacecrafts, majorly conducting experiments and maintaining spacecrafts, the astronaut profile slowly evolved to reflect the needs of the missions and different spacecrafts and space stations, as well as to fit the narrative that would best provide support to the space agencies.

Another grand step for mankind is about to happen, once more with a different objective and mindset—that of settlements and exploration of other planetary bodies. The ever-increasing autonomy from Earth that these missions require is a driving force for the case of parastronauts, as well as the mirror of human society in future settlements, integrating everyone for everyone. This is reflected in ESA's 2021 astronaut recruitment campaign, with, for the first time, the active recruitment of an astronaut with a lower limb mobility. Several NASA astronauts have also flown after cancer treatment and invasive surgery, showing that the health requirements are somewhat more flexible than one might believe. This chapter discusses the astronaut qualities and requirements, the specificities of Mars missions and why parastronauts could be the future of astronauts, and the steps to be taken to achieve inclusion of parastronauts into space missions.

2. Astronaut skills and requirements

Astronaut is perhaps some of the toughest job description and broadest skillset required on the job market, a job where almost every experience, hobby, and skill matter: mechanic, firefighter, glider pilot, medic, explorer, plumber hobbyist, and linguist. Astronauts have not just served to perform research in space but served for representation and a display of excellency, increasing the list of skills required on an already extremely demanding application. The astronaut qualities and requirements can be separated into the main frame—or the skills essential to the well-being of a mission and common to most space missions from the 1960s to today—and the era-dependent skills, which reflect the type of mission and geopolitical situations. An example of each would be 3D visual abilities for the main frame and social media outreach and communication for era-dependent skills.

While in the 1960s, the first missions required a pilot or test pilot background, the astronaut profile quickly shaped into a more versatile skillset, such as pilot and engineer, pilot and scientist, and then medical doctor or scientist. Indeed, as missions transitioned from mere orbital demonstrations to more complex tasks—Moon landings, experiments in microgravity, and early spacewalks—the astronauts' capabilities and training shifted. During the Apollo and early Skylab missions, the importance of engineering knowledge became clear. It was no longer enough to simply operate the craft; astronauts were expected to troubleshoot it. They became a blend of mechanic and operator, expected to function in contingencies where ground control support could be delayed or unavailable.

During the Space Shuttle era, civilians such as teachers and psychologists also became astronauts, popularizing access to space and, once more, changing the requirements expected of astronauts. Experiments onboard the shuttle required highly specific expertise, increasing the importance of payload compared to previous missions. This led to the recruitment of astronauts with backgrounds in medicine, geology, oceanography, and other sciences—some with little or no flying experience.

On the topic of flying and pilot skills, while at first, spacecraft required test pilot skills due to the complexity of operation and docking, the advent of computers drastically reduced the amount and complexity of manipulation required for spacecraft operation. In the 1960s, the spacecraft was an extension of the pilot, thus the emphasis on physical and psychological endurance was at its highest. Weightlessness, isolation, and confinement were novel human experiences, and there were very few baselines or historical precedents to guide selection processes. This trend continued through the Space Shuttle era, which marked a decisive turning point in the evolution of astronaut selection. The reusable nature of the shuttle and the increasing complexity of its payload meant astronauts were also scientists, teachers, and technicians. The expansion of the astronaut corps to include women and civilians reflected a shift in space exploration priorities.

As missions and technology evolved, the 90s and early 2000s saw that automatic docking would be of necessity for refueling of the ISS and, mostly, Mars missions where the crew would not be able to perform a reliable docking after the long trip, nor be present on Mars for further operations [1].

This explains how hard skills and background are shaped by technology and mission requirements. However, the other aspect of astronaut requirements is the ill- or well-fitting of the astronaut in the political landscape. During the Cold War era, for instance, the focus was placed on elite, high-performance individuals who could operate under intense pressure and represent national prestige. In contrast, the current astronaut profile is more focused toward teamwork, more collaborative and communicative, and oriented toward long-duration missions, scientific research, and international cooperation.

In the 2000s, the rise of international collaboration, especially the development of the International Space Station (ISS), cultural sensitivity, and language skills entered the picture. ISS crews are often multilingual and multinational, making interpersonal and intercultural communication a crucial component of mission success. The long-duration nature of ISS expeditions also highlighted the importance of group dynamics, emotional resilience, and psychological screening. Candidates now undergo months of intensive training not only in spacecraft systems and extravehicular activities (EVAs) but also in conflict resolution, stress management, and cross-cultural cooperation.

Today, the astronaut selection process is among the most rigorous in the world, drawing from thousands of candidates to find a small number of suitable individuals. An example of which is the 2021 ESA astronaut recruitment, with over 22,000 applicants, of which only 5 astronauts and 11 reserve astronauts were selected. Beyond the formal academic and physical criteria, space agencies increasingly emphasize behavioral competencies. Resilience, adaptability, and collaborative spirit are weighed just as heavily as cognitive aptitude or physical health. Astronauts are expected to be generalists in some areas—able to adapt to unfamiliar technical problems—and specialists in others, often conducting complex experiments in biology, chemistry, or physics with limited ground support.

Looking at the 2020s astronaut requirement for ESA and NASA, the following is listed [2, 3]:

1. Be a U.S.A. (NASA) or ESA-member state (ESA) citizen

2. Possess a master's degree in a STEM field, including engineering, biological science, physical science, computer science or mathematics, from an accredited institution.

3. Have at least 2 years of related professional experience obtained after degree completion or at least 1000 hours pilot-in-command time on jet aircraft.

4. Valid class 1 medical examination

In addition to these four points, the astronaut candidates must also have skills in leadership, teamwork, and communication. Typically, a quick select-out is performed by the agencies with medical screening as well as alcohol/drug screening, situational judgment tests, integrity/honesty tests, counterproductive work behavior, and personality tests. In the case of the European Space Agency, there is a 6-stage selection process, which processes everything from cognitive abilities and hard skills to teamwork and valid advanced medical examination. Personality and mental abilities are assessed in several ways, from general behavior to online testing and interviews. Some of the points assessed are the following:

- Mental/emotional stability

- Performance under stressful conditions

- Group-living skills

- Teamwork skills

- Family issues

Team exercises are typically a necessity in space missions due to high level of cooperation needed to operate spacecrafts and space stations. Team exercises provide an assessment of an applicant's ability to perform in a team as well as a subset of competencies.

These skills are required from astronauts working for space agencies and staying an extended period of time in space, but not requested to the same extent from private astronauts that have purchased tickets. The rise of commercial spaceflight and the advent of long-duration exploration missions—particularly those targeting the Moon and Mars—are again reshaping the skills landscape. Diving into those skills will show the wide pool of background and people, which is becoming the future of astronaut profiles.

Missions beyond low Earth orbit (LEO) will still require a very high level of skillset, mental stability, and training, as they pose new challenges: radiation exposure, deep-space communication delays, autonomy in life-critical decisions, and prolonged isolation. Crew members must be trained in medical procedures beyond basic first aid, sometimes including surgical assistance, due to the impossibility of rapid evacuation. The psychological toll of a multi-year mission is immense, and psychological screening now includes simulations of high-stress, long-duration isolation to assess mental fortitude.

Additionally, astronauts are now expected to act as very active ambassadors for space exploration, requiring strong public speaking and media skills. In an era of livestreams from space and public engagement through social media, the line between technical expert and science communicator has blurred. Astronauts must be able to vulgarize mission and space research for the general public and maintain a high level of public visibility and trust [4, 5].

The increase in versatility of missions, new prospects, and mission goals will most likely contribute to an expansion of definition of the skillsets of astronauts. With commercial space companies lowering the cost of access to space and missions becoming more routine (though still high-risk), new profiles are emerging. The increasing modularity and adaptability of spacecraft systems are reducing the barriers for non-pilot astronauts while also demanding greater adaptability and creative problem-solving under uncertainty.

As we now compare the skills and requirements placed for the astronaut selection, an additional factor comes to mind: what the candidates can control, and what they cannot. Looking at **Figure 1**, we can see that the hard skills and soft skills are placed together as skills that can be trained, skills that the astronaut candidates have control over, as they can be learned and developed. On the other side of the diagram, cognitive skills and health can only be controlled to a certain extent, while social background and nationality can rarely be changed. Typically, a congenital heart condition will bring to an immediate halt the candidate's application. As for nationality or background, an example is given with the European Space Agency's system, where each member country contributes differently. Countries with a higher investment in human space exploration will increase the chances of their candidates being selected as astronauts. In the same way, only US citizens can apply to NASA's astronaut call. This is not something that candidates have real control over, regardless of their suitability for the program.

These stringent requirements have contributed to an interest for commercial astronauts—for those who would not pass either or some of the tests or simply would never reach these requirements. In addition, the profession of astronauts necessitates years of basic training followed by several months of mission-specific training, health checks, and outreach assignments, something that does not fit every individual. The astronaut training is incredibly costly to space agencies, over several million euros per astronaut, thus it is in the interest of the agencies that the recruited and trained astronauts can be assigned to several missions and remain healthy during that course.

Thus, there is an increasing demand for missions to be accessible to all, with a reduced amount of skills or limited health, suitable for very short commercial

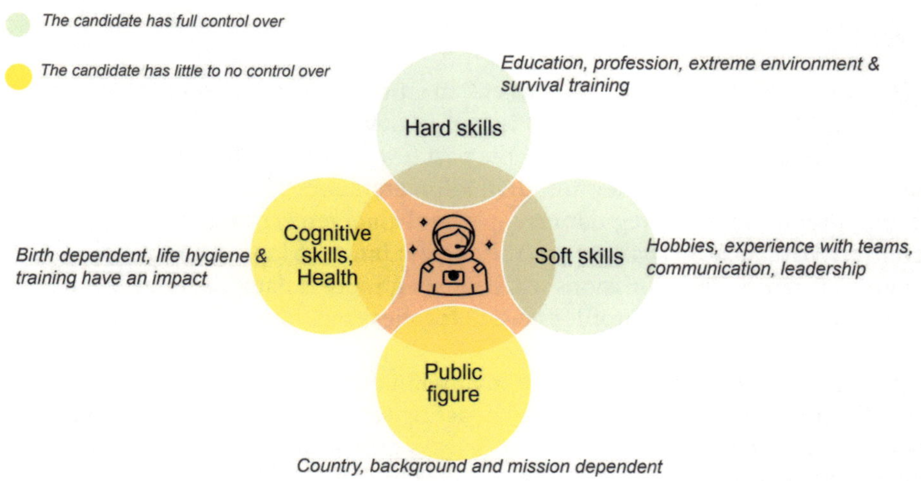

Figure 1.
Summary of the general astronaut skills.

missions. Health, in particular, can be redefined. While the European Space Agency put an age limit of 50 years old for the 2021 recruitment, astronauts and private astronauts over 70 have flown to space. NASA astronaut Donald Pettit returned to Earth on his 70th birthday, on April 20th 2025, after a 220-day stay on the International Space Station, and John Glenn returned to space with the space shuttle at the age of 77. Notably, Wally Funk, who had completed the training for Mercury 13 before the program was discarded, flew a suborbital flight in 2021 at the age of 82 with Blue Origin.

In other health considerations, astronauts Scott Kelly and Chris Hadfield have both flown before and after medical operations. Scott Kelly had been diagnosed with prostate cancer, while Chris Hadfield had to undergo abdominal surgery to remove scar tissue. In both cases, health considerations had to be taken into account—such as the decision to undergo an operation rather than radiation not to overshoot the radiation quota set for astronauts by NASA. Hayley Arceneaux became the first astronaut with a prosthetic leg to orbit the Earth, in 2021, having survived bone cancer as a child [6, 7].

This notably shows that medical screening can accommodate for old age, cancer, prosthetics, and even abdominal surgery. And it has been reflected in the 2021 ESA call for astronauts, where for the first time ever, astronauts with lower limb disabilities were recruited. A year and a half later, John Mcfall was announced as the first space agency astronaut with a lower limb deficiency, in this case, with an amputated leg above the knee. Since then, several calls for projects and opportunities have been released to prepare for his mission, and John Mcfall completed his qualification for the ISS in February 2025.

Some of these changes from agency astronauts to private astronaut recruitment medical requirements are already reflected in NASA's OCHMO-STD-100.1A. First released in 2021, the latest draft (2024) includes the requirements for private astronauts and agency astronauts. In addition, the medical assessment to be conducted preflight is mission duration dependent (respectively <30 days or > 30 days), as well as NASA Suborbital research specialists, showing the different categories of medical evaluation possible [8]. This is another representation of the current times changing toward a more open space for all.

3. Mars missions: The parastronaut's field

Moon and Mars space missions present an entirely different set of operational conditions and challenges compared to missions in Low Earth Orbit (LEO), which have largely defined human spaceflight for the past few decades. While the International Space Station (ISS) is only a few hundred kilometers above Earth and benefits from constant ground support, regular resupply, and emergency evacuation possibilities, Moon and Mars missions will stretch the current limits of logistics, autonomy, and human endurance. These missions will be longer, more isolated, and far less forgiving. As such, the astronaut profiles needed for these missions will likely require some adaptation.

One of the first key differences lies in mission duration and autonomy. A round-trip to the Moon is comparatively short—lasting several days or weeks—but a Mars mission may take between 18 months to 3 years depending on orbital mechanics and surface stay duration. This difference impacts not only the physical preparation of the crew but also their mental resilience and problem-solving capabilities. On Mars, communication delays of up to 22 minutes each way mean that real-time support from Earth is impossible. Real-time video conferencing, which is routine on the ISS, becomes impossible, as sending a message and receiving a response can take nearly an hour. This fundamentally changes the relationship between astronauts and mission control. Rather than being guided, crews must brief Earth retrospectively, reporting on decisions already made and actions already taken. Pre-scripted procedures and rigid decision trees will have to be supplemented—or replaced—by AI assistance and real human judgment, forged through training and experience. Astronauts will have to proceed to medical procedures, mechanical repairs, systems troubleshooting, and even operational leadership without mission control input and little help.

This will likely affect crew composition and training. For instance, a small crew might include a medical doctor, but all members will likely need cross-training in emergency medicine, diagnostics, and surgical support, to a much more advanced level than they would have for LEO missions. The same principle applies to engineering and system management. Each astronaut will need to be fluent in multiple domains and capable of thinking on their feet when systems fail, supplies are limited, and external help is unavailable.

Environmental conditions also influence skill requirements. The Moon has no atmosphere, extreme temperature changes, and lower gravity, while Mars, although having an atmosphere, features dust storms, lower gravity, and significant radiation exposure. The difference in terrain, gravity, and atmospheric composition requires astronauts to master new forms of EVA, pressurized suit usage, and mobility techniques. Astronauts will also need to handle complex habitat systems, life-support management, and in-situ resource utilization (ISRU), such as extracting water from Martian soil or producing oxygen on the Moon. These activities demand engineering, chemistry, and geology skills well beyond what was expected from previous crews.

Psychological resilience is also paramount. The extended isolation, distance from Earth, and limited team size all amplify interpersonal tensions and risks of mental fatigue. Astronauts for Moon and Mars missions must be chosen not only for their cognitive capacity but also for emotional intelligence, stability under extreme conditions, and aptitude for living in high-stress group settings. Because of the distance to Earth and the resulting autonomy, crew members may play dual roles—not only as mission crew but also as onboard mediators, as the mission control center will be able to offer little support. The psychological screening will include simulations of isolation and conflict resolution over months, not just weeks.

Furthermore, space settlements go beyond exploration; they represent the first steps toward human habitation outside Earth. As missions shift from scientific expeditions to long-term bases and possibly even permanent colonies, the skills required will change once more. No longer limited to STEM professionals, off-Earth settlements will require cooks, artists, educators, maintenance workers, IT support, and even legal and governance experts. Building a self-sustaining community on another planetary body means thinking beyond space agencies and considering humanity as a whole. If settlements are successful on the Moon and Mars, the duration of a human stay on Earth might span from a few months to a whole life, thus shifting the psyche and overall motivation and perspective of astronauts on planetary bodies.

This shift in astronaut skills can be represented once more with a bubble schematic. **Figure 2** shows an updated diagram of the skills that could be required of astronauts for Moon and Mars missions. In this, the public figure skillset has been replaced with adaptability—as the mindset to withstand the trip to Mars will have to change once the astronauts arrive on Mars and have to exploit skills in a completely new environment; as once they are on Mars, they will have to experience this high

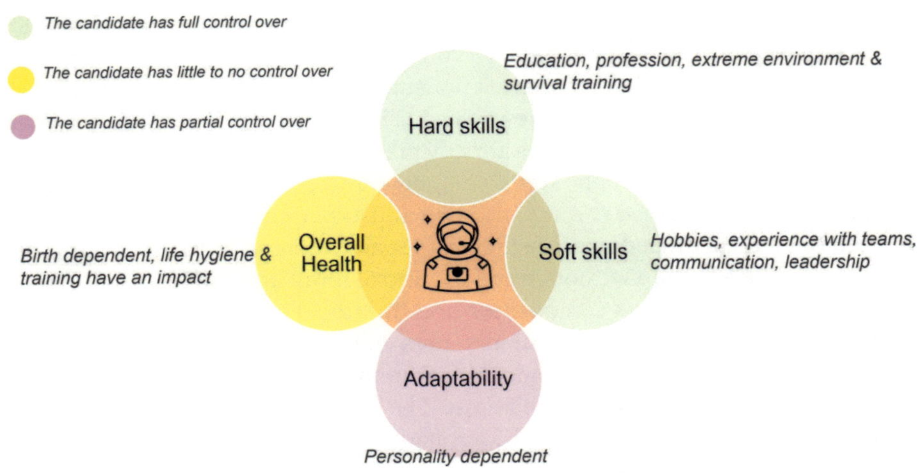

Figure 2.
Updated skill scheme for astronauts to Mars.

level of autonomy in terms of communication; and as on Mars, the possibility of accidents and subsequent handicaps arising will skyrocket. The distance and lack of direct communication between the Earth and Mars will most likely lead to a decrease in public interest or will have to be replaced with commitments from other areas of space agencies, as astronauts will not be able to provide as much on-site presence, thus freeing some space for candidates with higher adaptability. As was also previously discussed, missions have started to be more adapted for parastronauts and astronauts with previous health issues such as cancer. This is also reflected in the 'overall health' bubble of the schematic.

Indeed, the shift to adaptability would also correlate with the growth of the different astronaut skillset, background, and health levels, as well as life-course on Mars. The longest stay in space being 437 days (roughly 14 months), we do not possess data on the effects of microgravity for stays of 3 years or more. It is to be expected that astronauts' health will continue to degrade once they reach Mars, despite a return to partial gravity. With settlements, astronauts might spend 3, 10, 20 or more years, with potential long-lasting medical effects.

While parastronauts and astronauts alike can be screened on Earth for cardiovascular impairment, as well as organ damage or congenital diseases, nothing will make them injury-proof. An astronaut might become temporarily or permanently blind—an example of which was astronaut Chris Hadfield during an EVA, who was unable to see due to contamination in his eyes—they may break an arm or leg and require adapted protocols and systems. The inclusion of parastronauts in the astronaut cohort might provide onsite assistance to astronauts on Mars that might suffer from a debilitating accident. From teaching how to cope and adapt their movements to displays of empathy, parastronauts could lead and help manage the complexities of Mars missions. Empathy, teamwork, and communication will be skills of paramount necessity for such long missions, and parastronauts could provide better insight and coping mechanisms to the crew due to their own condition [9].

Another point that could possibly make parastronauts suitable for Mars missions is a result of the fluid shift within the human body in reduced gravity. On Earth, gravity sends blood toward our feet, and our heart pumps it back up, but on the international space station and any spacecraft, the reduced gravity or microgravity sends our blood rushing back to our heads. This is one of the reasons why our cardiovascular system is less challenged in space, as the heart must provide less effort to pump the blood back toward our head and torso, resulting in the need to train for 2 h per day aboard the ISS (including muscle and bone degradation). For parastronauts who would be missing lower limbs, this difference should technically be reduced, specifically in astronauts missing both legs. This could lead to a cardiovascular system that would be less prone to degradation and less bone degradation that could result in a number of nephrology issues such as kidney stones from calcium deposits [10]. In addition, this reduced fluid shift could perhaps also lessen optical degradation experience by astronauts. This difference in pressure gradient between short and tall astronauts can be based on the hemodynamic model set by Lawley et al. [11]. In their research, the vessel length as well as the astronaut characteristics such as tissue are used as factors to calculate the pressure through blood vessels in the human body. A higher tissue weight corresponds to higher intercranial pressure, a direct effect of fluid shift in microgravity. However, the long-term effects of reduced tissue weight via the absence of limbs still remains to be proven [12].

In terms of performances, Wiedmann et al. [13] have researched the importance of several physical performances preflight, during flight, and postflight. Their research, part of the 'parastronaut feasibility project,' was to investigate if physical performance tests should be part of future astronaut recruitments for an inclusive ESA astronaut corps to test their fight readiness. The study aimed to understand which areas of physical performance would have to be evaluated to mimic nominal and off-nominal performances in crew tasks and activities.

The study showed that the main physical skills to be examined were space-related, mission-specific coordination skills of the upper extremities. This was followed by endurance performance, stamina, dexterity of the upper extremities, and motor learning ability, as reflected in **Figure 3**. This has correlated with ESA's parastronaut recruitment, which only accepted lower limb deficiency, while upper limb mobility has so far been too high of an importance. It highlights that for the inclusion of

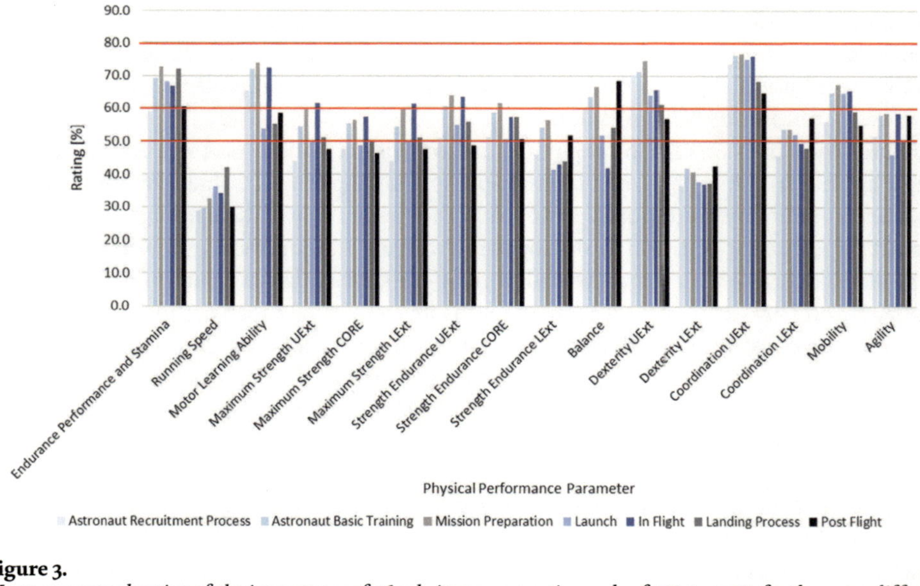

Figure 3.
Mean percentual rating of the importance of 16 sub-items on a rating scale of 0% to 100% for the seven different stages of an astronaut's career. The prioritization is assigned in three different levels: level 1 (very important) has a range of 61–80%, level 2 (moderately important) a range of 50–60%, and level 3 (negligible) a range for everything < 50%.

parastronauts with higher limb impairment, missions, systems and protocols must be adapted. The study did not research other disabilities such as sensory impairment.

Analogue space missions, space missions' simulations, have been conducted in 2017, 2023, and 2024 with crews that had analogue parastronauts. The mission duration varied between a few days and 2 weeks, and each crew had a parastronaut. The disabilities included: a blind woman, a blind man missing an arm and some fingers, a man missing a hand, an active wheelchair user, and a woman missing a leg [14]. These mission simulations have already started paving the way for parastronauts and modulable habitats, showcasing that a more inclusive and adaptive mission and habitat profits all crew members, regardless of disability, leading the way for Moon and Mars mission preparations by supporting test beds for space agencies.

4. Parastronauts: A timeline for the future of astronauts

While space mission simulations started as early as 2017 to include parastronauts, several organizations have started to push in this direction. Parabolic flights with wheelchair users (notably Steven Hawking) or blind or deaf astronauts, have already taken place. The rise of private missions and the qualification of astronaut John McFall in sprint 2025 for the ISS have also quick-started the adaptation of protocols, recruitment processes, and system autonomy, thanks to the Parastronaut Feasibility Project and now Fly! campaign [15, 16].

From the 1960s, the skills required for the piloting of spacecrafts have become less of a priority, thanks to technological advances. On the other hand, the need for mental resilience, coping mechanisms, empathy, and cross-functional skills has vastly increased, which are all skills that do not require 2 legs, 2 hands, or 2 arms, or any sort of specific limb function. As NASA is preparing for the return to the Moon in the coming years, China is operating its space station, and the Gateway is being planned; mission protocols and other systems can already start being modified to accommodate different disabilities for the Moon or Mars. Previous flights have shown that space can accommodate young and old, healthy and cancer survivors alike, with or without legs. Space is large enough to accommodate more parastronauts and pave the way to safer and more autonomous Mars missions.

While Moon missions are in preparation, Mars missions and Moon settlements are still in their infancy in terms of human life support systems, spacecrafts, habitats for human settlements, and higher autonomy protocols. While seemingly disheartening, this can be a perfect opportunity for the inclusion of parastronauts and injured astronaut protocols and modifications. LEO missions as well as analogue space missions can be used to train astronauts and parastronauts; measure empathy levels, coping mechanisms, and communication levels; study crew dynamics and performance; as well as compare both the mental and physical resilience of parastronauts. The effects of microgravity on parastronauts with shorter or absence of lower limbs will provide clarity on effects of microgravity on the human body, allowing for once the distinction between the effects of the blood shift toward the upper body and the reduced need for muscle uses.

With medical requirements being currently reworked, it is also the perfect timing to include different abilities and handicaps while not compromising requirements that would disqualify candidates, such as epilepsy, which can present immediate danger to crews and mission.

Another aspect of disabilities not covered is that of mental health disabilities. Mental health is a day-to-day taboo, but is of upmost taboo in space, and little information is to be found on the mental health of astronauts. While it is known that some slight depression and anxiety can arise during space missions, the extent is not commonly known. The longest space mission was only surpassed by the longest analogue space mission, the MARS520 mission, which confined astronauts in a simulated spacecraft aimed for a return trip to Mars. Upon reaching the last months of the mission, the crew experienced lower moods and depression, as isolation and communication were fully akin to Martian trips conditions. However, studies suggest that microgravity and radiation could also impact mental health and cognitive abilities, showing that the need for further research is very much needed [17, 18].

As astronauts in long-duration spaceflights may experience depression and lower mood, perhaps individuals with treated depression or other mental health issues or neuro-atypicality such as anxiety and ADHD, who have developed coping mechanisms on Earth, may be better suited to assist crew members who experience anxiety or depression for the first time. This hypothesis would need to be tested in remote analogue bases, such as in the Arctic before being tested in space, as its basis relies on the premise that high-functioning individuals in a stressful, high-responsibility environment are isolated from the rest of humanity. The presence of peers who have experienced similar symptoms in the past, and despite so have been selected and work alongside them, could perhaps assist astronauts in opening up about their mental health and symptoms, such as current trends show on Earth [19, 20].

However, in the parastronaut timeline, due to the taboo nature of mental health, it is unlikely that the focus of parastronaut inclusion will be targeted on this topic. In the future, however, it will be a crucial point to assess for the future of Moon and Mars mission settlements.

It is likely that the timeline for parastronaut summarizes as such:

1. First parastronaut analogue mission (completed – 2017)

2. First parastronaut selection

3. First suborbital parastronaut flight (completed 2021)

4. Parastronaut with lower limb deficiency qualification for ISS (2025)

5. First parastronaut ISS flight in the coming years

6. Recruitment of parastronaut with upper body deficiency or sensory deficiency

7. Concurrent modification of spacecrafts and mission protocols for new generations of parastronatus

8. Qualification of new parastronauts for space missions

Thus, we should expect John McFall's mission in the coming years to propel parastronaut research forward, in conjunction with analogue space missions and parabolic flights adding further testbeds for new protocols, technology, and research to the topic of parastronauts.

5. Conclusion

The astronaut profile, while being one of excellency, has been constantly shifting since its first definition. With a planned return to the Moon, Gateway and Mars mission, the time might be just right to propel parastronauts forward as new astronauts, whose disabilities might just be a super ability in space. The shift to including parastronauts implies a broader and more inclusive vision for space, one that is following that of space agencies around the world. The use of parabolic flights, analogue space missions and private space missions help further research and push back the limits of the human body, showcasing parastronaut capabilities, and paving a future on the Moon and Mars where astronauts, injured astronauts, parastronauts can work together in autonomy.

Acknowledgements

Thank you to LunAres Research Station for the leadership in starting parastronaut analogue space missions, and to my CHASM and PARACHASM team for the work conducted together.

Acronyms and abbreviations

ADHD	Attention Deficiency and Hyperactivity Disorder
EVA	Extra-Vehicular Activity
ISS	International Space Stations
ISRU	In-Space Resources Utilization
LEO	Low-Earth Orbit
STEM	Science Technology Engineering Maths

Author details

Eleonore Poli
CHASM, CSEM, Lausanne, Switzerland

*Address all correspondence to: eleonorec.poli@astro-chasm.org

References

[141] Polites ME. An Assessment of the Technology of Automated Rendezvous and Capture in Space. NASA TP-1998-208528. Alabama: Marshall Space Flight Center; 1998

[142] NASA. Become an Astronaut. Available from: https://www.nasa.gov/humans-in-space/astronauts/become-an-astronaut/ [Accessed: April 20, 2025]

[143] ESA. Astronaut Selection 2021-22 FAQs. https://www.esa.int/About_Us/Careers_at_ESA/ESA_Astronaut_Selection/Astronaut_selection_2021-22_FAQs#:~:text=ESA%20plans%20to%20recruit%204,created%20reserve%20pool%20of%20astronauts [Accessed: April 20, 2025]

[144] Bingaman J, Dawson W. Media use and US attitudes toward government and private space exploration. Space Policy. 2022;**62**

[145] Bertrand PJ, Niles SL, Newman DJ. Human spaceflight in social media: Promoting space exploration through twitter. New Space. 2015;**3**(2):117-133. DOI: 10.1089/space.2015.0004

[146] Hadfield C. An Astronaut's Guide to Life on Earth. Life Lessons from Space; 2013

[147] Kelly S. Enrudance: A Year in Space, a Lifetime of Discovery. Knopf Doubleday Publishing Group; 17 Oct 2017

[148] DRAFT NASA. Office of the Chief Health and Medical Officer, NASA Spaceflight Medical Selection Recertification and Mission Evaluation Standards; 2024

[149] Powell A, Myers T. Developing mental toughness: Lessons from paralympians. Frontiers in Psychology. 2017;**8**:9,10,14

[150] Debra G-M, Suleyman G. Numerical characterization of astronaut CaOx renal stone incidence rates to quantify in-flight and post-flight relative risk. npj Microgravity. 2022;**8**:14-15, Article 2

[151] Mohammadyari P, Gadda G, Taibi A. Modelling physiology of haemodynamic adaptation in short-term microgravity exposure and orthostatic stress on earth. Scientific Reports. 2021;**11**(1):4672. DOI: 10.1038/s41598-021-84197-7

[152] Lawley JS et al. Effect of gravity and microgravity on intracranial pressure. The Journal of Physiology. 2017;**595**(6):2115-2127. DOI: 10.1113/JP273557

[153] Wiedmann I, Weerts G, Brixius K. The ESA Parastronaut feasibility project: Investigating the need and contents of physical performance tests for an inclusive European astronaut corps. Sports Medicine. 2023;**53**(11)

[154] Poli E, Ducai T. Accessibility Study in Analogue Space Missions: ICARES-2 Missions and Control Group for Parastroauts, Presented at the International Astronautical Congress. Milan: Italy; 2024

[155] Chelsea G. Disabled Space Enthusiasts can Now Apply for Zero Gravity Space Training. Available from: https://www.space.com/astroaccess-zerog-flight-disabled-astronaut-training [Accessed: April 20, 2025]

[156] UK Space Agency. First Astronaut With a Disability Cleared for Space Station Mission. Available from: https://www.gov.uk/government/news/first-astronaut-with-a-disability-cleared-for-space-station-mission

[157] Marazziti D, Arone A, Ivaldi T, Kuts K, Loganovsky K. Space missions: Psychological and psychopathological issues. CNS Spectrums. 2022;**27**(5):536-540. DOI: 10.1017/S1092852921000535

[158] Arone A et al. The Burden of Space Exploration on the Mental Health of Astronauts: a Narrative Review; 2021

[159] Naslund JA, Aschbrenner KA, Marsch LA, Bartels SJ. The future of mental health care: Peer-to-peer support and social media. Epidemiology and Psychiatric Sciences. 2016;**25**(2):113-122. DOI: 10.1017/S2045796015001067

[160] Oddo L, Knause L, Safren S. Investigating resilience to depression in adults with ADHD. Journal of Attention Disorders. 2026;**22**(5):5, 6

Chapter 6

Traveling Companions: The Human Microbiome and the Future of Space Exploration

Douglas Hamilton and Gabrielle Caswell

Abstract

The human microbiome has evolved in tandem with humans for hundreds of thousands of years, shaped by diet, lifestyle, and environment. As space agencies and commercial ventures set their sights on Mars and Exploration Class Space Missions (ECSM), traveling beyond Earth's biosphere and into biologically closed, artificial environments will present new challenges for maintaining human health, microbiota and microbiome stability, and life support systems as this complex symbiotic relationship faces unprecedented disruption. This chapter explores the concept of biomes: Earth's naturally evolved environments of essential communities and life forms, which have evolved over 2.5 billion years, and macrobiomes, ecosystems established in isolated, closed environments such as submarines, the International Space Station (ISS), and eventually, missions such as those planned for Mars. In space, microbiomes have evolved in response to confined habitats. Studies from ISS crew members have shown microbial profiles of astronauts and their environment interacts, creating a dynamic feedback loop between crew members' microbiome, the built habitat, and their diet. The human microbiome is susceptible to environmental and dietary changes, leading to gastrointestinal dysbiosis, including those predicted to occur during long-duration spaceflight: altered gravity, cosmic radiation, circadian rhythm disruption, and enclosed habitat are linked to immune suppression, metabolic disturbances, neurocognitive deficits, and perhaps, with longer exposure, disease progression, and transcriptome anomalies. This manuscript proposes that the success of interplanetary colonization hinges on our ability to understand, preserve, and possibly enhance the complex interplay between human biology, the microbiota and microbiome, and artificial habitats and ecosystems.

Keywords: macrobiome, microbiome, Phys Chem systems, Mars, ECSM, human health

1. Introduction

Life on Earth began with the smallest and most enduring form of life: microorganisms. As Earth's original pioneers, they appeared some 2.5 billion years ago, laying the foundations for all complex life forms [1]. They appeared in the first oceans and were only anaerobic heterotrophic bacteria because the primordial atmosphere was almost

IntechOpen

entirely devoid of oxygen. Cyanobacteria evolved at least 2.4 billion years ago. They became Earth's first photosynthesizers, making food using carbon dioxide, water, and the Sun's energy, and releasing oxygen as an end by-product. Photosynthesis triggered a chain of planetary transformations that made Earth habitable. The cyclic nature of oxygen production and cyanobacteria's demise and the reversal of this process continued for over a billion years. Eventually, it resulted in the oxidation and reduction of iron in the seas, and oxygen-dependent life forms flourished. Over billions of years, they evolved into astonishingly diverse life forms: bacteria, viruses, archaea, fungi, and algae. These microbes adapted and flourished in the oceans, deserts, mountaintops, and deepest geothermal vents. Every corner of the planet teems with microscopic life, as does every human being [2–6].

Today, we understand that humans are not solitary organisms but living ecosystems. Our bodies host vast communities of bacteria, viruses, fungi, and archaea collectively known as the microbiome. This term refers not only to the microbes but also to the entire set of microbial genes. The microbiome has deeply integrated functions, helping humans digest food, extract nutrients, support immunity, regulate inflammation, maintain overall health and physiological functioning, and mental health [7]. In adults, microbiome composition is influenced by environmental pH, location, role and utility of the body part, race, familial genetics, and dietary intake. Three main habitats have been defined on the human body: the skin and its appendages, the oral-gastrointestinal tract, and, in females, the reproductive tract. Although physically located within us, these microbiome habitats are technically outside the human body.

The symbiotic relationship between humans and their microbiome is not static; it changes in response to our environment, diet, stress, hygiene, and even the company we keep [7–10]. With the emergence of modern humans, estimated to have occurred approximately 120,000 years ago, humans migrated from ancient hunter-gatherer societies to agricultural and then industrial civilizations. The microbiome adapted, shifting in composition and diversity in response to our changing diets, housing, and hygiene practices. It evolves alongside us, shaped by what we eat, where we live, and how we interact with others and the natural world; it is the unseen half of our biology.

The microbiome remained relatively stable until the advent of agriculture, approximately 12,000 years ago, then adapted to barley and corn, and was subsequently modified to accommodate wheat and rice crops. The development of small settlements and the rise of cities, craft specialization, and more compact urbanized housing and shelter shifted food production outside towns and settlements [7]. Accompanying hygiene, food scarcity, and the development of food storage and preservation techniques impacted the microbiome in two ways: reducing the diversity of the gastrointestinal microbiome and then removing the positive impact of nature on the immune system [11, 12].

With the advent of World War I in the early twentieth century, new food preservation technologies extended the life of food, albeit at the cost of an accompanying decrease in microbiota; with some preservation methods, sterile food could last up to 2–3 years. Urban living, processed foods, and sterilized environments reduce the diversity of the microbiome, a change that has been linked to rising rates of obesity, allergies, and autoimmune diseases [11–13].

This delicate relationship is challenged in enclosed environments such as submarines, space stations, or a future Mars habitat, all of which create a sealed environment for their crews, forming a closed ecological system, a macrobiome [14]. Studies on the International Space Station have shown that the microbiome of astronauts shifts in response to spaceflight, as do the microbial communities on the station itself [13, 15]. The people aboard, and the microorganisms they carry, interact with that craft environment in ways we are only beginning to elucidate; new microbial communities have emerged, with new patterns of behavior and physically shaped by confinement, radiation, altered gravity, and recycled air and water [16–20]. Over time, the ISS surfaces reflect the microbial makeup of the crew living there, forming a new, self-contained ecosystem [9, 21–23]. As we prepare for multi-year journeys to Mars and beyond, we must understand how to maintain a healthy microbiome in space, how the macrobiome in space alters its function, and how to design habitat macrobiomes that support life at every level, from the tiniest microbe to the human explorers themselves.

This raises profound questions. If we build homes on Mars, how do we create habitats for the life forms that sustain us? [21] Can we engineer supportive ecosystems in space that keep humans alive and help us thrive? [22, 24] What happens to the microbiota and microbiome under the conditions of the macrobiome, or artificial life support? And how do we prepare ourselves, biologically and ethically, for the unintended consequences of bringing Earth's biology to another planet? More than passengers, these microorganisms are collaborators, and in the story of humanity's expansion into the cosmos, they are among our most trusted, biologically essential, traveling companions [25–27].

2. The human microbiome

The microbiome is one of our most quietly influential companions in the vast, interconnected systems that make up the human body. It is an extensive ecosystem of trillions of bacteria, viruses, fungi, and other microscopic life forms that live on and inside us. These microorganisms, collectively known as our microbiome, are not just passengers; they are collaborators [13]. They help us digest food, regulate our immune systems, and influence our mood, cognition, and mental health [11, 20, 22, 28, 29].

When the balance is disrupted, a condition known as dysbiosis can occur [22, 30, 31]. Dysbiosis is more than just a scientific term; it represents a shift from health toward disease [25, 32–38]. It can weaken the gut barrier, increase inflammation, and is associated with metabolic disorders, autoimmune diseases, and mood changes. In space, where immune function is altered, the risks posed by a disturbed microbiome are magnified [19, 32, 39].

Evidence from ground-based simulations and spaceflight missions, such as NASA's MARS500 experiment and studies aboard the International Space Station (ISS), reveals that although astronauts' microbiomes are remarkably resilient, they change composition within the gut during missions [13, 23, 39, 40]. The human microbiome is expected to play a critical role in astronaut health during long-duration space missions. Space is an extreme environment marked by microgravity, cosmic radiation, confinement, recycled air, absence of fresh food, altered day-night cycles, and limited microbial exposure from natural environments, influencing the diversity and composition of the gut and skin microbiota [10, 12, 13, 32, 40–42].

Studies found that astronauts' skin microbiomes are similar to the surfaces touched on the ISS, rather than their microbiomes before flight [13, 22, 43]. In turn, the surfaces of the ISS began to reflect the unique microbial fingerprint of the crew, creating a shared, enclosed ecosystem, or what could be called the "spacecraft macrobiome." [9, 15, 23, 43, 44] Other research has shown that skin health can deteriorate during long missions, presenting symptoms like dryness, rashes, and irritation, which is not just uncomfortable; it is a sign of profound disruptions in microbial and immune functioning of the skin [22, 43, 45]. Fluctuations in certain bacteria have been linked to changes in immune responses and inflammation, raising questions about long-term health impacts during multi-year space journeys [26, 32, 41–43]. A striking finding is that the gastrointestinal microbiomes of astronauts tend to become more alike during a mission, partly due to their shared diet [15, 22, 43].

Understanding how the human microbiome responds to spaceflight is not just about avoiding rashes or stomachaches [45]. This understanding is an appreciation of the human body's first line of defense, maintaining immune and physiological resilience in the face of unfamiliar stressors, and ensuring long-term adaptability as we stretch the boundaries of a human presence beyond Earth [33, 34, 44, 46].

2.1 The effect of the space environment on living organisms

Space is not empty. It is filled with invisible forces, microgravity, radiation, and isolation that test every living system, including the human body and its microscopic passengers [19, 30, 31, 46]. As we prepare to send humans on long missions to destinations like Mars, we must confront how these forces impact our bodies, our cells, our genes, and the ecosystems we carry with us [40, 43, 47–51].

2.2 The transcriptome in microgravity

DNA is the genetic instruction manual of life at the heart of every cell [52–55]. But not every instruction is always used [55]. Instead, cells read from the DNA as needed, creating messenger RNA (mRNA) transcripts that instruct the cell on which proteins to produce [24, 55, 56]. This collection of active instructions is referred to as the transcriptome, which can shift in response to the environment, acting like a dynamic expression of who we are at a molecular level [24, 57].

Radiation-induced damage can especially harm organs with high cell turnover, such as bone marrow and the intestinal lining [19, 30]. In space, this expression changes; radiation can damage DNA, prompting some cells to shut down, stop dividing, or self-destruct through a process called apoptosis, or programmed cell death [24–26, 45, 46, 56]. This is a natural safeguard, but it also means that tissues that rely on rapid renewal, like the skin, gut lining, or blood cells, may become compromised [19, 22, 32, 46]. These effects have been observed in lab animals and modeled for human physiology, underscoring the importance of shielding and mitigation strategies on future missions [27, 31, 58–60].

2.3 Replication changes in space

Some tissues in our bodies replace themselves at astonishing rates. For instance, the small intestine's lining can regenerate completely every few days [30]. In humans, it is estimated that we replace roughly 100 billion intestinal cells daily. In the microgravity of space, this regeneration process is disrupted, the forces that guide cell division and migration are altered, and radiation may introduce mutations along the way [22, 32, 46, 61, 62]. For cells that are supposed to grow, divide, and die in a tightly controlled rhythm, space represents a kind of biological confounder. This disruption is not just a cellular issue; it could have real consequences for digestion, nutrient absorption, immune function, and the integrity of the gastrointestinal barrier, a crucial line of defense against pathogens [22].

In short, space changes us at the molecular level; our bodies are incredibly adaptive, but those adaptations come at a cost, which may rise the longer we stay in space [15, 22, 25, 30, 32, 60, 62, 63]. As we plan for years-long missions, understanding how our genes, cells, and tissues respond to these challenges will be critical [14, 32, 64, 65]. The key to thriving beyond Earth may lie in engineering solutions and decoding how life reshapes under alien conditions [15].

3. Traveling companions

As we look toward Mars and other worlds, we must confront a profound truth: humans never travel alone; we are ecosystems, living, breathing collections of countless microscopic companions [13, 23, 32, 39, 42, 43]. These microbes, and the relationships they maintain with each other and with us, are just as vital to our survival in space as oxygen and water [40, 42]. The success of a long-duration mission may depend as much on maintaining this invisible support crew as it does on rocket thrust or life support systems [15, 44, 57].

3.1 Earth's biome: A blueprint for Mars

Earth's biome is the product of 2.5 billion years of evolution; it is a vast, interlinked system where microbes, plants, animals, and humans live in complex harmony [12, 13]. This living system has weathered mass extinctions, ice ages, and planetary collisions and emerged resilient [1]. When we send a crew to Mars, we are not just sending humans; we are exporting a carefully selected subset of Earth's biome. But which organisms should we bring? Should we include insects to manage waste or bacteria to help recycle nutrients? Could we use fungi to support soil ecosystems? What roles might plants, animals, or even marine life play?

And what if this biome, compressed into a habitat, isolated from Earth's vast ecosystems, evolves in unexpected ways under Martian gravity and radiation? [48, 49] Will tomatoes still grow the same way? Will they be the same after 50 seed-to-seed harvests? Will beneficial bacteria stay friendly? [22] These are not just scientific questions; they are survival questions.

3.2 Microbial communication and quorum sensing

Back on Earth, microbes communicate and cooperate using chemical signals in a process called quorum sensing [19]. Through these signals, bacteria can determine the number of their own kind nearby and collectively decide when to activate specific genes, such as those involved in forming biofilms or producing antibiotics [19, 32]. In space, these behaviors change. Bacteria in microgravity grow thicker cell walls and sometimes form unusual structures, including "outer membrane vesicles"—tiny packages used to communicate with other microbes [19, 25, 32, 46, 63].

These changes could affect how microbes behave during missions, shifting the balance between helpful and harmful species. The longer a mission lasts, the more likely the spacecraft's microbial ecosystem will evolve away from its Earth-bound origin [16, 17, 32, 39, 53, 54, 66]. That shift could influence everything from digestion to disease risk, and we must be prepared to monitor and adapt accordingly [24].

3.3 Building a closed, regenerative life support system

If we are to live on Mars, we must build more than a spacecraft—we must create an Earth-like bubble capable of sustaining life for years at a time. A closed, bioregenerative life support system (LSS) aims to recycle air, water, food, and waste using living organisms, thereby reducing our dependence on supplies from Earth [26]. But this is no small feat; the system must be reliable over many generations of plants, bacteria, and potentially animals [19, 26]. We must understand whether algae, fungi, or vegetables grown in Martian gravity will behave the same way they do on Earth after 50 or 100 generations. Will their genetic stability hold? Will their role in the system remain predictable?

We cannot calculate such systems' risk or resilience without confidence in these answers [26]. We also cannot ignore the possibility of mutation: life is under stress in space, and change happens fast. The mean time between failure (MTBF), a measure of how long a system runs before breaking, becomes a critical concept for machines and the living systems we rely on.

Essentially, our microbial passengers are not just along for the ride but co-architects of survival. To build a livable Mars, we need more than engineering; we need to cultivate life, not just our own but the tiny, ancient, tireless allies that have shaped Earth and now may help us shape a new home.

3.4 Mars mission life support

Surviving on Mars is not just about getting there. It is about staying alive, safely, healthily, and productively for months or even years. This will require more than airtight suits and recycled water [11, 32, 42, 63]. It will demand an entirely new kind of life support system that works in harmony with biology, not just technology.

3.5 The macrobiome of a Mars mission

When astronauts travel to Mars, they bring more than themselves; they bring an entire macrobiome, the total of all living systems that travel with them. This includes their microbiomes (bacteria, viruses, and fungi) as well as intentionally transported organisms, such as food plants, algae, beneficial microbes, and potentially small animals. All of these organisms must survive and function in an artificial habitat far removed from Earth. And just as importantly, they must support the astronauts' activities of daily living, such as eating, bathing, exercising, maintaining hygiene, and staying mentally healthy [26, 45]. The confined space of a spacecraft or Martian habitat means that human systems and the habitat macrobiome are tightly intertwined, causing the line between crew and their environment to blur [15, 22, 23, 42–44]. Everything, from microbes in our gut to bacteria on our skin, becomes part of a delicate, shared ecosystem. In effect, the human body is part of the life support system.

3.6 The human microbiome: A genetic powerhouse

Our microbiome, the trillions of microbes living within and on our bodies, contains up to 100 times more genes than our DNA [11, 15, 64, 67]. These microbial genes influence digestion, immunity, inflammation, mood, and cognition. Still, neither the human genome nor the microbiome has ever been exposed to a Mars-like macrobiome, recycled air and water, and total separation from Earth's ecosystems. We do not yet know how these living systems will adapt—or what could go wrong over months and years [22].

Will friendly microbes become aggressive? Will essential bacteria die off? What if an unnoticed shift in the gut microbiome triggers inflammation, weakens the immune system, or impairs mental health? These are no longer theoretical concerns—they are mission risks.

3.7 The human genome in a new world

When the human genome was sequenced in 2001, many hoped it would be the ultimate manual for life [64, 68, 69]. The reality has proven far more complex; DNA does not act alone; it responds to environmental cues, such as stress, diet, or radiation, by turning specific genes on or off [24, 34, 45, 52–55, 57, 58, 70, 71]. In space, these environmental cues radically differ from those experienced on Earth; solar and cosmic radiation, for example, can damage DNA, trigger mutations, or increase cancer risk [20, 67, 72–74]. Some scientists even propose that RNA, the messenger molecule between DNA and proteins, may act as a bridge between the environment and the genome, allowing for real-time editing of responses [15, 24, 59, 61, 62, 75].

In such a dynamic system, the question becomes: How will genes behave in a Martian environment? And how might our microbiome and genome adapt—or fail to adapt—over generations?

4. Physical-chemical life support systems (PhysChem)

Current space missions utilize open-loop non-regenerative life support systems based on physical and chemical processes, often called "PhysChem," to recycle air and water [26, 45]. These systems are predictable, reliable, and easy to maintain; however, they heavily depend on supplies from Earth, which is unsustainable for long-term missions or colonies [26]. New PhysChem systems that recover water from sweat, urine, and exhaled breath are being tested. This recovered water can then be split into hydrogen and oxygen, with the hydrogen recombined with carbon dioxide to produce more water and methane. These closed-loop chemical processes are clever but not enough on their own [19, 26].

4.1 The reliability challenge

In engineering, we measure system reliability using the mean time between failures (MTBF), the average time an item operates before it breaks. We always engineer machines, but we must now also integrate biology for complex systems like a regenerative life support system. Let us build a bio-based macrobiome using living organisms like algae to produce oxygen or utilize microbes to process waste. We must know: will it still work after many generations of replication under Martian gravity? What if radiation alters the behavior of those organisms? What if a key microbe mutates or dies off?

Without a backup system, the consequences of failure could be catastrophic. That is why most mission planners propose combining a biological life support system with a fail-safe PhysChem backup to ensure redundancy and survivability.

In summary, creating a life support system for Mars is not just about recycling air or water; it is about building a living, breathing ecosystem that includes humans, their microbiota, microbiomes, and a macrobiome, all comprised of a host of carefully selected companions. The challenge is not just technological; it is biological, genetic, and ecological.

4.2 Mars mission profiles

Sending humans to Mars is not a one-size-fits-all journey. Mars expeditions could follow very different profiles depending on the mission design, technology readiness, and risk acceptance. Whether a short visit or a multi-year settlement, each version presents unique challenges, and all of them revolve around one key question: how do we keep humans alive and healthy so far from Earth?

4.3 Earth-Mars transit

Traveling from Earth to Mars could take up to 3 months. During this time, astronauts would rely on familiar life support systems, like those used aboard the International Space Station (ISS). These open-loop systems are well-tested and reliable for relatively short durations, recycling water and air and managing waste, but they still require non-regenerative expendable supplies.

The real challenge begins after arrival. Mars missions may involve surface stays of 18–36 months or longer, too extended to rely on resupply. Shipping massive quantities of air, water, and food to Mars is unsustainable. Instead, crews must begin producing and recycling these resources locally.

This is where a biologically regenerative life support system comes in—what we refer to as a "closed-loop macrobiome." However, such systems require incubation and may not function perfectly immediately.

One proposed solution is a hybrid approach: begin the mission with a reliable PhysChem system to provide immediate life support, allowing time for the biological systems—the macrobiome—to establish themselves. This strategy could provide a safety net if the macrobiome does not function as planned or becomes unhealthy and requires "treatment."

Another consideration is mission length. With current life support technology, a shorter stay on Mars—perhaps only 1 month—might be feasible, primarily if the mission uses a "gravity-assist" route that slingshots past Venus to reach Mars. This shortcut has trade-offs, including more time spent in deep space and higher radiation exposure, as the transit brings the spacecraft closer to the Sun. Nevertheless, it may reduce the risk of long-term biological system failure.

4.4 The goal: A closed-loop life support system

The holy grail of Mars exploration is a closed-loop, biologically regenerative habitat where plants, microbes, and humans form a sustainable, interdependent ecosystem. This system would produce oxygen, recycle waste, and grow food, minimizing dependence on Earth.

But building this kind of life support system is extraordinarily difficult. It has to work reliably, not just for weeks or months, but for years—possibly even decades. The ecosystem, from algae to tomatoes to beneficial bacteria, would need to survive and reproduce in Mars's reduced gravity and higher radiation environment. What happens to these organisms after 10, 50, or 100 generations in space? We do not yet know.

DOI: http://dx.doi.org/10.5772/intechopen.1011344

That is why many experts suggest deploying these biological systems before the crew arrives. Pre-deploying greenhouses, compost systems, and microbial colonies would allow the macrobiome to stabilize, creating a healthy, living environment ready for human arrival. The backup is a PhysChem system that can take over temporarily as the biological system performs a "reboot'" if something goes wrong. In a remote Martian outpost, help is not on the way. The illness or death of a macrobiome life support system due to mutation, contamination, or unexpected interaction with the crew may be lethal, so redundancy is key. An alternative macrobiome embodiment must be available to deploy and assume the role as the primary macrobiome.

4.5 Mission-critical design questions

Designing a sustainable life support system for Mars means confronting tough questions:

- What organisms should we bring from Earth? Just plants and microbes, or small animals too?
- How do we prevent ecosystem collapse? Could a single mutation compromise the system's homeostasis?
- Can we predict how organisms will behave after years of space radiation and Martian gravity?
- What if the microbial community becomes unbalanced and starts damaging the system?

These questions do not just affect mission planning—they impact everything from crew health to the viability of colonization. They underscore the need for continued terrestrial testing and verification in analog environments, simulations, and perhaps pre-Mars mission simulations on the Moon.

In essence, a successful Mars mission will depend on the creation and survival of a resilient, adaptable, and sustainable life support system. Whether it is a short visit or a multi-year colony, the future of Mars exploration may hinge on our ability to engineer not just machines but robust and reliable microbial-based ecosystems.

4.6 Mars mission crew selection

Sending humans to Mars is not just about rockets and robots—it is about people. And choosing the right crew for a multi-year mission in deep space may be one of the most difficult and consequential decisions ever made in space exploration. It is not just about who is strong, intelligent, and trained—it is about who can thrive in a strange new world, far from Earth, in an environment that will challenge every cell in their body.

4.7 Matching biology with the mission

A Mars crew will live in a completely artificial macrobiome environment, which may impact medical wellness from bone strength to immune function to mental health [20]. That is why future crew selection might extend beyond psychological screening and fitness tests [20, 59, 60, 64, 75].

We may eventually use genomic and biological profiling to understand how individual astronauts are likely to respond to life on Mars [60]. Advances in "omics", genomics (genes), transcriptomics (environmental gene expression), proteomics (proteins and structure), and metabolomics (organism energy) could reveal who is most resilient to the space environment [59, 64, 75, 76].

Will some people have biological advantages? [20, 64, 67, 77, 78] Some genes might make a person more resistant to radiation, better at processing low oxygen levels, or less likely to develop osteoporosis in low gravity. Understanding these traits could help ensure not only mission success but also survival.

5. Ethics and privacy

But this raises complex ethical questions. If a person's genome suggests a higher risk of cancer or other illnesses, should they be disqualified? [61, 64, 79] What about someone with a family history of inherited disease or a predisposition that may never become a problem? [36–38, 67, 80]

Who has the right to see this information, and who gets to decide based upon it? [79] Could insurers, governments, or private companies misuse genetic data? These questions are already being debated here on Earth and will become even more urgent as we plan for Mars [81].

Balancing privacy, fairness, and safety will be among the thorniest challenges in preparing for deep space missions. This is not just a medical or legal issue; it is a profoundly human issue.

5.1 Exploring beyond low earth orbit

Once astronauts leave Earth's immediate neighborhood beyond the protective bubble of low Earth orbit (LEO), they will be on their own. Communications will lag, supplies will not arrive, and every system, including the human body, will be stressed.

Technological advances in wellness, such as personalized medicine and genomic diagnostics, may help manage health in space [64, 75]. However, utilizing these technologies requires understanding the interactions between the crew's biology and the macrobiome they inhabit, including the microbiome surrounding and within them [15, 34, 59, 82]. Space agencies and private space companies will need to navigate this new territory carefully, advancing science while upholding the rights and dignity of every crew member [20].

5.2 What we have learned so far

The NASA Twins Study, in which one twin spent a year on the International Space Station [20]. At the same time, the other remained on Earth, providing an unprecedented glimpse into how space affects the human body. The results were precise: spaceflight changes gene expression, immune function, and cognitive performance [20].

In recent years, civilian missions—like SpaceX's Inspiration4—have collected massive amounts of biological data [61]. Thousands of samples from these short flights are helping researchers build a "space health atlas" that will guide the design of future missions and crew care [60, 65]. Still, long-duration space missions are rare; only a handful of people have spent more than a year in space. We know little about how space affects women during pregnancy, children during development, or people as they age. Yet these are precisely the kinds of people who will 1 day form the backbone of a Mars colony.

5.3 Can we predict health in space?

No one can predict precisely how a person will respond to the extreme conditions of a multi-year Mars mission. Will a dormant genetic condition become active under stress? Will a normally harmless gut microbe turn dangerous in low gravity? Could the skin microbiome shift in a way that triggers chronic irritation or inflammation?

We also have to consider practical risks, such as the probability of someone developing pancreatitis, a stroke, or cancer far from Earth [36–38, 80]. Would they be treatable? Could the rest of the crew carry on the mission? These questions are not just medical—they are ethical, emotional, and operational. At the same time, some genetic traits might prove to be beneficial. For instance, mutations that increase bone density or help the body cope with low oxygen might be mission-enabling [59, 64, 75].

5.4 Living together with a shared microbiome

There's another, subtler factor in crew compatibility: microbiomes. Each person carries a unique ecosystem of bacteria, viruses, and fungi, particularly in their gut and on their skin [15, 22, 42, 65]. When people live in close quarters, their microbiomes begin to blend and interact with one another. This matters because the microbiome influences immune function, mood, digestion, and other aspects of health [14, 20]. Some experts suggest that Mars crews should live together for a period before the mission begins—to let their microbiomes adapt to each other, potentially reducing health risks once they are in space.

These ideas, which mix science, ecology, and sociology, represent a new way of thinking about astronaut health: not as individuals but as interdependent parts of a shared, living system.

5.5 Commercial deep space missions and colonization

The race to Mars is no longer just a government venture. Private companies and space agencies worldwide are preparing to send humans beyond low Earth orbit—to the Moon, Mars, and potentially beyond. The first long-duration deep space missions, Exploration Class Deep Space Missions (ECDSMs), will be stepping stones to permanent human settlements. But building a new society on another planet requires more than rockets—it demands biology, resilience, and foresight.

5.6 Building life on the Moon and Mars

Colonizing Mars means bringing life with us—not just astronauts, but also plants, animals, and microorganisms essential to survival. These organisms form part of a macrobiome—a miniature, closed version of Earth's living systems that must function reliably in a Martian environment.

But how do we decide which life forms to bring? Do we include insects to help recycle waste? Should we rely on bacteria and fungi to enrich the soil? Will marine life play a role in food production or water purification? We must consider the ecosystem as a whole, not just individual species, but the intricate relationships that make life on Earth work.

This is a monumental task. Earth's biosphere is the result of billions of years of evolution and adaptation. Mars is cold, dry, and bathed in radiation. Can Earth's biology survive—and thrive—there?

6. The genetic challenge

6.1 Another question lurks beneath the surface: What kind of human population should we send to Mars?

If only a small group of colonists makes the trip, their genetic diversity will be limited. Over generations, this could lead to a phenomenon known as the founder effect, where rare genetic traits become common simply because the population started from a limited gene pool. On Earth, this has contributed to certain inherited diseases in isolated populations, such as Tay-Sachs among Ashkenazi Jews and Ellis–van Creveld syndrome in the Amish.

Genetic diversity will be key to a thriving future colony. A wide range of alleles (gene variants) offers protection against diseases and enhances the population's ability to adapt. However, managing that diversity, especially if missions select for specific traits such as radiation resistance or bone density, may pose serious ethical dilemmas.

6.2 The microbiome on Mars

As important as our genes are, our microbiomes may be just as critical. These vast communities of bacteria, viruses, and fungi live on our skin, gut, and the macrobiome environment. They play a crucial role in helping us digest food, fighting disease, and regulating our mood.

Once we arrive on Mars, our microbiomes will be cut off from Earth's. In closed habitats, they may begin to evolve in new and unpredictable ways—potentially beneficial or potentially harmful. For example, the effect of Martian dust on our internal ecosystems remains unknown.

Over generations, these changes could create a new kind of human whose biology is shaped not just by DNA but also by their altered microbiomes from life on another planet.

6.3 The human experiment

Mars colonization is not just a technical challenge—it is a massive human experiment. We are taking a slice of Earth's biology and dropping it into an alien world. Every aspect of life—from how we breathe and eat to how we grow food and raise children—will be tested.

The microbiome will play a central role. We are only beginning to understand how deeply it influences human health, immunity, and mental function. These influences may become even more pronounced on Mars in closed systems with recycled air and artificial diets.

Humans' phenotypic plasticity—or adaptability—will be stretched to its limits. How far can we bend before we break? Will we evolve into something new?

6.4 Beyond the resume: Rethinking crew selection for Mars

Sending humans to Mars is not just about choosing the fittest or the smartest—it is about selecting individuals whose bodies, minds, and microbiomes can adapt and thrive in the unknown. The criteria for choosing a Mars crew go far beyond the traditional astronaut checklist of physical health, technical skill, and psychological resilience. In deep space, our biology becomes part of the spacecraft, and the selection process becomes more scientific and philosophical.

7. Risk, health, and the unknown

Simulated Mars missions on Earth—such as NASA's HERA (Human Exploration Research Analog) and the year-long Crew Health and Performance Exploration Analog (CHAPEA) mission in a 3D-printed Martian habitat—have taught us a great deal about isolation, teamwork, and daily life in confined environments. However, they have also revealed how much we still do not know.

What happens when a crew member suffers a serious illness or injury millions of kilometers from home? What if a medical emergency is linked to a unique shift in the microbiome in space? How does one crew member's microbiome affect another's? These questions are not theoretical but biological realities we must face before launch.

Some experts suggest that pre-mission cohabitation could help crews adjust to each other's microbiomes before entering the spacecraft. The idea is to reduce microbial "shock," align their ecosystems, and foster resilience. This slight shift could have significant consequences for crew health and performance.

8. Microbiome: The invisible teammate

The microbiome is not just a collection of microbes—it is a dynamic part of the human system. It influences digestion, immunity, cognition, and even emotion. In space, this silent partner could make or break the mission.

As astronauts adjust to new environments, diets, and sleep cycles, their microbiomes shift too [77]. A well-balanced microbiome may help regulate stress and inflammation, while a disrupted one can trigger a range of issues, including gastrointestinal problems, skin rashes, and impaired cognitive function [14, 20, 22].

The gut-brain axis, which links the gastrointestinal system to brain function, is an emerging area of research with powerful implications. Understanding how microgravity, isolation, and synthetic food environments affect this connection is crucial in deep space, where psychological health is already under pressure.

We are learning that mitochondria—the energy-producing engines of our cells—may also be deeply influenced by the microbiome. This means microbial health might impact cellular health, physical endurance, and aging. These relationships must be understood, preserved, and engineered for long-term spaceflight and planetary colonization.

9. Personalizing crew biology

As our understanding of genetics and personalized medicine advances, a question arises: Should astronauts be selected based on their genetic makeup?

Some genetic traits may be mission-enabling, like variants that improve bone density or reduce oxygen needs. Others may pose higher risks, like mutations linked to certain cancers or diseases that could be triggered by radiation or stress. However, drawing these lines is an ethically complex task.

What happens if an astronaut has a genetic risk for a disease that might never manifest on Earth but could be accelerated by space conditions? Should they be excluded? Could some alleles, once considered dangerous, become advantageous in space?

We do not yet have all the answers and must tread carefully. Genetics is only one part of the story. The interaction between genes, microbiomes, and the Martian macrobiome will be unpredictable. Ultimately, crew selection may not be about choosing the "perfect" humans but those with the most significant biological adaptability [70, 83, 84].

10. Conclusion: A new biology for a new world

For more than 2.5 billion years, Earth has been our only home. Its intricate web of life—its biome—has shaped every cell in our bodies. We carry that legacy within us: in our DNA, microbiomes, and microscopic ecosystems that inhabit and reside within us. As we prepare to leave this evolutionary cradle, we face a profound question: Can we take Earth's biology with us?

We'll need more than fuel, food, and oxygen to thrive on Mars or anywhere beyond low Earth orbit. In all its complexity, we'll need life, not just human life, but the microbial companions that digest our food, modulate our immune systems, and even influence our minds [14]. We'll need to create a macrobiome, a complete, functioning biological system, inside a spacecraft and Martian habitats. This is not just an engineering challenge; it is a challenge to our biological imagination. Unlike machines, ecosystems are dynamic; they mutate, adapt, and evolve, more so when under pressure. Radiation, altered gravity, synthetic macrobiomes, and time will transform our microbiomes, genes, and health in ways we are only beginning to understand. What evolves on Mars may not be what we launched from Earth.

These changes are not just risks; they are opportunities. They invite us to reimagine what it means to be human, not in isolation, but in symbiosis with the living world we bring along. They challenge us to create systems that are not just resilient but regenerative—to build habitats that keep us alive and help us adapt and thrive.

To do this well, we must embrace a new kind of exploration that sees biology as something to protect and partner with. We must select our crews carefully, not only for their strength and skills but also for their biological harmony. We must design our missions to support life as an integrated system, with feedback loops between humans, their microbiomes, and the macrobiome they inhabit.

The future of space exploration may not depend on any one technology or gene. It may depend on how well we understand our living selves—and how wisely we bring that life into the void.

In this sense, the journey to Mars is not just about reaching another planet. It is about bringing Earth with us, in spirit and substance. Our microbiomes are not just passengers on this voyage; they are traveling companions, co-architects of our survival, and perhaps, our evolution.

DOI: http://dx.doi.org/10.5772/intechopen.1011344

Acknowledgements

The authors would like to express their deep gratitude to the many scientists, clinicians, engineers, and explorers whose work, both on Earth and in orbit, has laid the foundation for our growing understanding of the human microbiome in space. We are especially thankful to the astronaut corps and their support teams, who have courageously participated in studies that bring science closer to the stars.

We are grateful to our microbiology, aerospace medicine, and environmental systems engineering colleagues for their insights and ongoing collaboration. Their expertise has been invaluable in shaping this multidisciplinary exploration of life beyond Earth.

We also acknowledge the support of academic institutions and space agencies that continue to champion the integration of biological science in space exploration and whose missions reflect a shared commitment to the future of life in space.

Finally, we recognize the inspiration drawn from countless past, present, and future individuals who see space as a destination and a canvas upon which to paint a new vision for humanity.

Conflict of interest

The authors declare that they have no conflict of interest.

Author details

Douglas Hamilton[1*] and Gabrielle Caswell[2]

1 Department of Internal Medicine, University of Calgary, Canada

2 Spaceport Australia Pty Ltd, Moree NSW, Australia

*Address all correspondence to: doctorhami@gmail.com

References

[161] Ostrander CM, Heard AW, Shu Y, et al. Onset of coupled atmosphere–ocean oxygenation 2.3 billion years ago. Nature. 2024;**631**:335-339. DOI: 10.1038/s41586-024-07551-5

[162] Thorogood A, Dalpé G, Knoppers BM. Return of individual genomic research results: Are laws and policies keeping step? European Journal of Human Genetics. 2019;**27**(4):535-546. DOI: 10.1038/s41431-018-0311-3

[163] Clinton WJ. Executive order 13145 of February 8, 2000: To prohibit discrimination in federal employment based on genetic information. Federal Register. 2000;**65**(28):6877-6880

[164] Voge VM, Rosado NR, Contiguglia JJ. Sickle cell anemia trait in the military aircrew population: A report from the military aviation safety Subcommittee of the Aviation Safety Committee, AsMA. Aviation, Space, and Environmental Medicine. 1991;**62**(11):1099-1102

[165] Milburn NJ, Neitz J, Chidester T, Lemelin M. New genetic technology may help pilots, aviation employees, and color vision researchers. Aviation, Space, and Environmental Medicine. 2013;**84**(11):1218-1220. DOI: 10.3357/asem.3860.2013

[166] Siew K, Nestler KA, Nelson C, D'Ambrosio V, Zhong C, Li Z, et al. Cosmic kidney disease: An integrated pan-omic, physiological and morphological study into spaceflight-induced renal dysfunction. Nature Communications. 2024;**15**(1):4923. DOI: 10.1038/s41467-024-49212-1

[167] Flies EJ, Clarke LJ, Brook BW, Jones P. Urbanisation reduces the abundance and diversity of airborne microbes - but what does that mean for our health? A systematic review. Science of The Total Environment. 2020;**738**:140337. DOI: 10.1016/j.scitotenv.2020.140337 Epub 2020 Jun 22

[168] Nastasi N, Haines SR, Bope A, Meyer ME, Horack JM, Dannemiller KC. Fungal diversity differences in the indoor dust microbiome from built environments on earth and in space. Scientific Reports. 2024;**14**(1):11858. DOI: 10.1038/s41598-024-62191-z

[169] Mora M, Perras A, Alekhova TA, Wink L, Krause R, Aleksandrova A, et al. Resilient microorganisms in dust samples of the international Space Station-survival of the adaptation specialists. Microbiome. 2016;**4**(1):65. DOI: 10.1186/s40168-016-0217-7

[170] Mora M, Mahnert A, Koskinen K, Pausan MR, Oberauner-Wappis L, Krause R, et al. Microorganisms in confined habitats: Microbial monitoring and control of intensive care units, operating rooms, cleanrooms and the International Space Station. Frontiers in Microbiology. 2016;7:1573. DOI: 10.3389/fmic

[171] Ryu EP, Gautam Y, Proctor DM, Bhandari D, Tandukar S, Gupta M, et al. Nepali oral microbiomes reflect a gradient of lifestyles from traditional to industrialized. bioRxiv [Preprint]. 3 Jul 2024:2024.07.01.601557. DOI: 10.1101/2024.07.01.601557. Update in: Microbiome. 2024 Nov 4;12(1):228. DOI: 10.1186/s40168-024-01941-7

[172] Selway CA, Mills JG, Weinstein P, Skelly C, Yadav S, Lowe A, et al. Transfer of environmental microbes to the skin and respiratory tract of

humans after urban green space exposure. Environment International. 2020;**145**:106084. DOI: 10.1016/j.envint.2020.106084. Epub 2020 Sep 22

[173] Potter JD, Brooks C, Donovan G, Cunningham C, Douwes J. A perspective on green, blue, and grey spaces, biodiversity, microbiota, and human health. Science of The Total Environment. 2023;**892**:164772. DOI: 10.1016/j.scitotenv.2023.164772. Epub 2023 Jun 10

[174] Chancellor JC, Blue RS, Cengel KA, Auñón-Chancellor SM, Rubins KH, Katzgraber HG, et al. Limitations in predicting the space radiation health risk for exploration astronauts. npj Microgravity. 2018;**4**(1):8. DOI: 10.1038/s41526-018-0043-2

[175] Voorhies AA, Mark Ott C, Mehta S, Pierson DL, Crucian BE, Feiveson A, et al. Study of the impact of long-duration space missions at the international Space Station on the astronaut microbiome. Scientific Reports. 2019;**9**(1):9911. DOI: 10.1038/s41598-019-46303-8

[176] Alekseev VR, Hwang JS, Levinskikh MA. Effect of space flight factor on dormant stages in aquatic organisms: A review of international Space Station and terrestrial experiments. Life (Basel). 29 Dec 2021;**12**(1):47. DOI: 10.3390/life12010047

[177] Simonsen LC, Nealy JE, Townsend LW, Wilson JW. Space radiation dose estimates on the surface of Mars. Journal of Spacecraft and Rockets. 1990;**27**(4):353-354. DOI: 10.2514/3.26149

[178] Simonsen LC, Nealy JE, Townsend LW, Wilson JW. Martian regolith as space radiation shielding. Journal of Spacecraft and Rockets. 1991;**28**(1):7-8. DOI: 10.2514/3.26201

[179] Potten CS. A comprehensive study of the radiobiological response of the murine (BDF1) small intestine. International Journal of Radiation Biology. 1990;**58**(6):925-973. DOI: 10.1080/09553009014552281

[180] Garrett-Bakelman FE, Darshi M, Green SJ, Gur RC, Lin L, Macias BR, et al. The NASA twins study: A multidimensional analysis of a year-long human spaceflight. Science. 2019;**364**(6436). DOI: 10.1126/science.aau8650

[181] NASA. NASA'S Crew Health and Performance Exploration Analog & Mars Dune Alpha Habitat Houston, Texas. 2024. Available from: https://www.nasa.gov/humans-in-space/chapea/habitat/

[182] Caswell G, Eshelby B. Skin microbiome considerations for long haul space flights. Frontiers in Cell and Developmental Biology. 8 Sep 2022;**10**:956432. DOI: 10.3389/fcell.2022.956432

[183] Avila-Herrera A, Thissen J, Urbaniak C, Be NA, Smith DJ, Karouia F, et al. Crewmember microbiome may influence microbial composition of ISS habitable surfaces. PLoS One. 2020;**15**(4):e0231838. DOI: 10.1371/journal.pone.0231838

[184] Destefanis E, Avşar G, Groza P, Romitelli A, Torrini S, Pir P, et al. A mark of disease: How mRNA modifications shape genetic and acquired pathologies. RNA. 2021;**27**(4):367-389. DOI: 10.1261/rna.077271.120

[185] Zea L, Larsen M, Estante F, Qvortrup K, Moeller R, Dias de Oliveira S, et al. Phenotypic changes exhibited by *E. coli* cultured in space. Frontiers in Microbiology. 28 Aug 2017;**8**:1598. DOI: 10.3389/fmicb.2017.01598

[186] Divya Harika P, Mehta KH, Pulluri SS, Rana P, Rajani H, Aiman A. Oral health in zero gravity: A comprehensive review of orofacial effects and countermeasures in spaceflights. Cureus. 2023;**15**(11):e49035. DOI: 10.7759/cureus.49035

[187] Knisz J, Eckert R, Gieg LM, Koerdt A, Lee JS, Silva ER, et al. Microbiologically influenced corrosion—More than just microorganisms. FEMS Microbiology Reviews. 2023;**47**(5):fuad041. DOI: 10.1093/femsre/fuad041

[188] Basner M, Savitt A, Moore TM, Port AM, McGuire S, Ecker AJ, et al. Development and validation of the cognition test battery for spaceflight. Aerospace Medicine and Human Performance. 2015;**86**(11):942-952. DOI: 10.3357/AMHP.4343.2015

[189] Roy-O'Reilly M, Mulavara A, Williams T. A review of alterations to the brain during spaceflight and the potential relevance to crew in long-duration space exploration. npj Microgravity. 2021;**7**(1):5

[190] Siddiqui R, Qaisar R, Goswami N, Khan NA, Elmoselhi A. Effect of microgravity environment on gut microbiome and angiogenesis. Life (Basel). 24 Sep 2021;**11**(10):1008. DOI: 10.3390/life11101008

[191] Mayhew TM, Myklebust R, Whybrow A, Jenkins R. Epithelial integrity, cell death and cell loss in mammalian small intestine. Histology and Histopathology. 1999;**14**(1):257-267. DOI: 10.14670/HH-14.257

[192] Cervantes JL, Hong BY. Dysbiosis and immune dysregulation in outer space. International Reviews of Immunology. 2016;**35**(1):67-82. DOI: 10.3109/08830185.2015.1027821

[193] Porteus M. Genome editing: A new approach to human therapeutics. Annual Review of Pharmacology and Toxicology. 2016;**56**:163-190. DOI: 10.1146/annurev-pharmtox-010814-124454

[194] Li H, Yang Y, Hong W, Huang M, Wu M, Zhao X. Applications of genome editing technology in the targeted therapy of human diseases: Mechanisms, advances and prospects. Signal Transduction and Targeted Therapy. 2020;**5**(1):1. DOI: 10.1038/s41392-019-0089-y

[195] Ferrucci L, Fabbri E. Inflammageing: Chronic inflammation in ageing, cardiovascular disease, and frailty. Nature Reviews. Cardiology. 2018;**15**(9):505-522. DOI: 10.1038/s41569-018-0064-2

[196] Forsmark CE, Vege SS, Wilcox CM. Acute Pancreatitis. The New England Journal of Medicine. 2016;**375**(20):1972-1981. DOI: 10.1056/NEJMra1505202

[197] Howes N, Greenhalf W, Stocken DD, Neoptolemos JP. Cationic trypsinogen mutations and pancreatitis. Gastroenterology Clinics of North America. 2004;**33**(4):767-787. DOI: 10.1016/j.gtc.2004.07.003

[198] Cohn JA, Neoptolemos JP, Feng J, Yan J, Jiang Z, Greenhalf W, et al. Increased risk of idiopathic chronic pancreatitis in cystic fibrosis carriers. Human Mutation. 2005;**26**(4):303-307. DOI: 10.1002/humu.20232

[199] Liu Z, Luo G, Du R, Sun W, Li J, Lan H, et al. Effects of spaceflight on the composition and function of the human gut microbiota. Gut Microbes. 2020;**11**(4):807-819. DOI: 10.1080/19490976.2019.1710091

[200] Schroeder BO, Bäckhed F. Signals from the gut microbiota to distant

organs in physiology and disease. Nature Medicine. 2016;**22**(10):1079-1089. DOI: 10.1038/nm.4185

[201] Candela M, Biagi E, Turroni S, Maccaferri S, Figini P, Brigidi P. Dynamic efficiency of the human intestinal microbiota. Critical Reviews in Microbiology. 2015;**41**(2):165-171. DOI: 10.3109/1040841X.2013.813900

[202] Song SJ, Lauber C, Costello EK, Lozupone CA, Humphrey G, Berg-Lyons D, et al. Cohabiting family members share microbiota with one another and with their dogs. eLife. 2013;**2**:e00458. DOI: 10.7554/eLife.00458

[203] Turroni S, Rampelli S, Biagi E, Consolandi C, Severgnini M, Peano C, et al. Temporal dynamics of the gut microbiota in people sharing a confined environment, a 520-day ground-based space simulation, MARS500. Microbiome. 2017;**5**(1):39. DOI: 10.1186/s40168-017-0256-8

[204] Farkas Á, Farkas G. Effects of spaceflight on human skin. Skin Pharmacology and Physiology. 2021;**34**(5):239-245. DOI: 10.1159/000515963

[205] Santos MA, Correa LDF, Heberlé G. Health and hygiene of skin, hair, nails, and teeth in the space environment: Daily challenges. In: Pathak Y, Araújo dos Santos M, Zea L, editors. Handbook of Space Pharmaceuticals. Cham: Springer International Publishing; 2020. pp. 1-22

[206] Little JB. Principal cellular and tissue effects of radiation. In: Kufe DW, Pollock RE, Weichselbaum RR, editors. Cancer Medicine. 6th ed. Hamilton, Ontario: Holland-Frei; 2003

[207] Kramer LA, Sargsyan AE, Hasan KM, Polk JD, Hamilton DR. Emissary veins and neuroanatomic changes in space. Radiology. 2013;**266**(1):363

[208] Hamilton DR, Sargsyan AE, Garcia K, Ebert DJ, Whitson PA, Feiveson AH, et al. Cardiac and vascular responses to thigh cuffs and respiratory maneuvers on crewmembers of the International Space Station. Journal of Applied Physiology (Bethesda, MD: 1985). 2012;**112**(3):454-462. DOI: 10.1152/japplphysiol.00557.2011

[209] Jones JA, Sargsyan AE, Barr YR, Melton S, Hamilton DR, Dulchavsky SA, et al. Diagnostic ultrasound at MACH 20: Retroperitoneal and pelvic imaging in space. Ultrasound in Medicine and Biology. Jul 2009;**35**(7):1059-1067. DOI: 10.1016/j.ultrasmedbio.2009.01.002. Epub 2009 May 7

[210] Sargsyan AE, Hamilton DR, Jones JA, Melton S, Whitson PA, Kirkpatrick AW, et al. FAST at MACH 20: Clinical ultrasound aboard the international Space Station. The Journal of Trauma. 2005;**58**(1):35-39. DOI: 10.1097/01.ta.0000145083.47032.78

[211] NASA. Mission Success: HERA Crew Successfully Completes 45-Day Simulated Journey to Mars: NASA; 2024. Available from: https://www.nasa.gov/centers-and-facilities/johnson/mission-success-hera-crew-successfully-completes-45-day-simulated-journey-to-mars/

[212] Lander ES, Linton LM, Birren B, Nusbaum C, Zody MC, Baldwin J, et al. Initial sequencing and analysis of the human genome. Nature. 2001;**409**(6822):860-921. DOI: 10.1038/35057062

[213] Coyne J. Famous Physiologist Embarrasses himself by Claiming that the Modern Theory of Evolution is in Tatters 2013. Available from: https://

whyevolutionistrue.com/2013/08/25/famous-physiologist-embarrasses-himself-by-claiming-that-the-modern-theory-of-evolution-is-in-tatters/

[214] Noble D. Reply from Denis Noble: What did Alan Hodgkin know? The Journal of Physiology. 2023;**601**(7):1313-1314. DOI: 10.1113/JP284530

[215] Biémont C, Vieira C. Genetics: Junk DNA as an evolutionary force. Nature. 2006;**443**(7111):521-524. DOI: 10.1038/443521a

[216] Ratcliff J, Simmonds P. Potential APOBEC-mediated RNA editing of the genomes of SARS-CoV-2 and other coronaviruses and its impact on their longer term evolution. Virology. 2021;**556**:62-72. DOI: 10.1016/j.virol.2020.12.018

[217] Christofi T, Zaravinos A. RNA editing in the forefront of epitranscriptomics and human health. Journal of Translational Medicine. 2019;**17**(1):319. DOI: 10.1186/s12967-019-2071-4

[218] Song X, Zhang G, Zhou Y, Li W. Behaviors and mechanisms of microbially-induced corrosion in metal-based water supply pipelines: A review. Science of The Total Environment. 2023;**895**:165034. DOI: 10.1016/j.scitotenv.2023.165034

[219] Schmidt MA, Goodwin TJ, Pelligra R. Incorporation of omics analyses into artificial gravity research for space exploration countermeasure development. Metabolomics. 2016;**12**:36. DOI: 10.1007/s11306-015-0942-0. Epub 2016 Jan 20

[220] Overbey EG, Ryon K, Kim J, Tierney BT, Klotz R, Ortiz V, et al. Collection of biospecimens from the inspiration4 mission establishes the standards for the space omics and medical atlas (SOMA). Nature Communications. 11 Jun 2024;**15**(1):4964. DOI: 10.1038/s41467-024-48806-z

[221] Jones CW, Overbey EG, Lacombe J, Ecker AJ, Meydan C, Ryon K, et al. Molecular and physiological changes in the SpaceX Inspiration4 civilian crew. Nature. 2024;**632**(8027):1155-1164. DOI: 10.1038/s41586-024-07648-x

[222] Grigorev K, Nelson TM, Overbey EG, Houerbi N, Kim J, Najjar D, et al. Direct RNA sequencing of astronaut blood reveals spaceflight-associated m6A increases and hematopoietic transcriptional responses. Nature Communications. 2024;**15**(1):4950. DOI: 10.1038/s41467-024-48929-3

[223] Edemekong PF, Bomgaars DL, Sukumaran S, Schoo C. Activities of daily living. In: StatPearls. Florida, USA: StatPearls Publishing LLC; 2024

[224] Overbey EG, Kim J, Tierney BT, Park J, Houerbi N, Lucaci AG, et al. The space omics and medical atlas (SOMA) and International Astronaut Biobank. Nature. 2024;**632**(8027):1145-1154. DOI: 10.1038/s41586-024-07639-y

[225] Rutter L, Barker R, Bezdan D, Cope H, Costes SV, Degoricija L, et al. A new era for space life science: International Standards for Space Omics Processing. Patterns (N Y). 2020;**1**(9):100148. DOI: 10.1016/j.patter.2020.100148

[226] Arora S, Puri S, Bhambri N. A designer diet layout for astronauts using a microbiome mediated approach. FEMS Microbiology Letters. 28 Jul 2022;**369**(1):fnac049. DOI: 10.1093/femsle/fnac049

[227] Afshinnekoo E, Scott RT, MacKay MJ, Pariset E, Cekanaviciute E,

Barker R, et al. Fundamental biological features of spaceflight: Advancing the field to enable deep-space exploration. Cell. 2020;**183**(5):1162-1184. DOI: 10.1016/j.cell.2020.10.050

[228] Nurk S, Koren S, Rhie A, Rautiainen M, Bzikadze AV, Mikheenko A, et al. The complete sequence of a human genome. Science. 2022;**376**(6588):44-53. DOI: 10.1126/science.abj6987

[229] McGuire AL, Caulfield T, Cho MK. Research ethics and the challenge of whole-genome sequencing. Nature Reviews. Genetics. 2008;**9**(2):152-156. DOI: 10.1038/nrg2302

[230] Wagner I, Braun M, Slenzka K, Posten C. Photobioreactors in life support systems. Advances in Biochemical Engineering/Biotechnology. 2016;**153**:143-184. DOI: 10.1007/10_2015_327

[231] Rutter LA, MacKay MJ, Cope H, Szewczyk NJ, Kim J, Overbey E, et al. Protective alleles and precision healthcare in crewed spaceflight. Nature Communications. 2024;**15**(1):6158. DOI: 10.1038/s41467-024-49423-6

[232] Camera A, Tabetah M, Castañeda V, Kim J, Galsinh AS, Haro-Vinueza A, et al. Aging and putative frailty biomarkers are altered by spaceflight. Scientific Reports. 2024;**14**(1):13098. DOI: 10.1038/s41598-024-57948-5

[233] Fried LP, Cohen AA, Xue QL, Walston J, Bandeen-Roche K, Varadhan R. The physical frailty syndrome as a transition from homeostatic symphony to cacophony. Nature Aging. 2021;**1**(1):36-46. DOI: 10.1038/s43587-020-00017-z

[234] Nascimento CM, Ingles M, Salvador-Pascual A, Cominetti MR, Gomez-Cabrera MC, Viña J. Sarcopenia, frailty and their prevention by exercise. Free Radical Biology & Medicine. 2019;**132**:42-49. DOI: 10.1016/j.freeradbiomed.2018.08.035. Epub 2018 Aug 31

[235] Ray S, Gebre S, Fogle H, Berrios DC, Tran PB, Galazka JM, et al. GeneLab: Omics database for spaceflight experiments. Bioinformatics. 2018;**35**(10):1753-1759. DOI: 10.1093/bioinformatics/bty884

[236] Schork NJ. Personalized medicine: Time for one-person trials. Nature. 2015;**520**(7549):609-611. DOI: 10.1038/520609a

[237] Nasrini J, Hermosillo E, Dinges DF, Moore TM, Gur RC, Basner M. Cognitive performance during confinement and sleep restriction in NASA's Human Exploration Research Analog (HERA). Frontiers in Physiology. 2020;**11**:394. DOI: 10.3389/fphys.2020.00394

[238] Oberg J. Does Mars need women? Russians say no. National Broadcast Company; 2005. Available from: https://www.nbcnews.com/id/wbna6955149

[239] Reed RD, Antonsen EL. Should NASA collect astronauts' genetic information for occupational surveillance and research? AMA Journal of Ethics. 2018;**20**(9):E849-E856. DOI: 10.1001/amajethics.2018.849

[240] Kuchenbaecker KB, Hopper JL, Barnes DR, Phillips KA, Mooij TM, Roos-Blom MJ, et al. Risks of breast, ovarian, and contralateral breast cancer for BRCA1 and BRCA2 mutation carriers. Journal of the American Medical Association. 2017;**317**(23):2402-2416. DOI: 10.1001/jama.2017.7112

[241] Genetic Information Nondiscrimination Act of 2008, Pub. L. No. 3 Stat. No. 110-233, 122 Stat. 881 (to be codified in scattered sections of 26, 29, and 42 U.S.C.) (Jan 2009). 2009

[242] Hamburg MA, Collins FS. The path to personalized medicine. The New England Journal of Medicine. 2010;**363**(4):301-304. DOI: 10.1056/NEJMp1006304

[243] Vogt C, Monai M, Kramer GJ, et al. The renaissance of the Sabatier reaction and its applications on earth and in space. Nature Catalysis. 2019;**2**:188-197. DOI: 10.1038/s41929-019-0244-4

[244] Fahrion J, Mastroleo F, Dussap CG, Leys N. Use of photobioreactors in regenerative life support systems for human space exploration. Frontiers in Microbiology. 2021;**12**:699525. DOI: 10.3389/fmicb.2021.699525

Chapter 7

The Psychological Needs of Future Astronauts

Francesco Pagnini and Francesca Grosso

Abstract

Exploring outer space presents profound challenges for human beings. Astronauts face isolation, confinement, microgravity, communication delays, and multiple stressors that can compromise psychological well-being and cognitive functioning. These challenges will intensify as missions become longer and shift beyond low Earth orbit toward the Moon and Mars. This chapter examines key psychological countermeasures designed to support astronauts in extreme environments. Automated psychological support systems, such as virtual counseling platforms and adaptive mood-monitoring software, offer immediate, tailored assistance, addressing mental health needs in real-time without relying on Earth-based communication. Mindfulness and relaxation techniques, including guided meditation protocols, breathing exercises, and progressive muscle relaxation, have enhanced stress resilience, emotional stability, and sustained cognitive performance in isolated settings like Antarctic stations and submarine crews. Structured routines that incorporate regular exercise, virtual reality environments for mood enhancement, and creative leisure activities further contribute to mental balance.

Keywords: astronauts, psychological well-being, microgravity, stress resilience, mindfulness

1. Introduction

The exploration of space stands as one of humanity's most daring and fascinating pursuits, an endeavor that pushes the limits of science, technology, and our ability to adapt. Beyond the technical and logistical challenges, however, long-duration space missions raise important questions about the psychological well-being of astronauts. These individuals, selected for their resilience, expertise, and teamwork skills, must operate in an extreme and unfamiliar environment characterized by isolation, confinement, microgravity, communication delays, and a range of other stressors that can impact their mental health and cognitive performance [1, 2]. As missions grow longer and more ambitious, with plans targeting the Moon, Mars, and beyond, it becomes increasingly important to understand astronauts' psychological needs and to design effective countermeasures to reduce the negative impact of the space environment on their well-being [3]. Establishing structured daily routines that include regular physical exercise, virtual reality (VR) environments to enhance mood, creative recreational activities, and mindfulness practices plays a crucial role in helping astronauts maintain psychological balance and motivation while creating a more engaging and supportive living environment. This chapter explores the key psychological challenges astronauts face during extended space missions, examining their causes, underlying mechanisms, and potential consequences. It also presents an integrated and personalized approach to support mental health and optimize performance, drawing on the latest research in space psychology, current scientific evidence, and lessons learned from both past and ongoing missions.

While this chapter focuses on the psychological challenges of astronauts involved in long-duration and deep space missions, it is important to recognize that psychological needs are relevant across all types of space participants, including governmental astronauts, commercial astronauts, and space tourists. Although long-duration missions amplify psychological stressors, even short-term missions such as suborbital flights, low Earth orbit missions, and lunar operations present unique psychological demands, such as acute stress, adaptation to novelty, management of expectations, and potential for performance anxiety [1]. Governmental astronauts, often engaged in scientific research and complex operational tasks, face stressors related to mission objectives and maintaining peak performance. Commercial astronauts may experience different pressures related to the success of commercial ventures and the safety of paying customers [4]. Space tourists, while undertaking shorter missions, require psychological support to manage the novel environment, potential disorientation, and the emotional impact of spaceflight [5]. Clarifying these distinctions is crucial to avoid the misconception that psychological countermeasures are only necessary for deep space missions. Investments in space psychology benefit all mission types and participants, ensuring safety, well-being, and mission success regardless of the destination or duration. Furthermore, the psychological strategies and interventions developed for spaceflight have significant relevance and applicability to terrestrial settings, as will be discussed later in this chapter.

2. How isolation and confinement affect the mind and social bonds

Isolation and confinement represent one of the most pervasive and complex psychological challenges for astronauts. The lack of social interaction beyond the crew, the distance from family and friends, limited living space, and the monotony of the environment can lead to feelings of loneliness, boredom, depression, anxiety, and irritability. Studies conducted in similar isolated settings, such as Antarctic research stations, submarines, and offshore oil platforms, have shown that prolonged isolation can negatively affect mood, motivation, sleep quality, cognitive function, immune system performance, and overall mental health [6]. In addition, isolation and confinement can distort the perception of time and space, creating a sense of disconnection from reality and increased vulnerability to stress. Isolation can also impact interpersonal relationships within the crew, increasing the risk of conflict, misunderstandings, and communication difficulties [7]. In such situations, implementing strategies that promote social and psychological support is vital to strengthening crew cohesion, ensuring effective communication, resolving conflicts, and fostering a positive, collaborative environment. Given that humans are inherently social beings, the lack of meaningful interactions can have a significant impact on psychological well-being. Microgravity, a defining feature of the space environment, presents a significant challenge to astronauts' sensory and motor systems [8]. In the absence of gravity, the human body undergoes a series of physiological and neurological changes that can affect spatial perception, balance, motor coordination, and orientation. In the short term, microgravity can cause spatial disorientation, nausea, vomiting, headaches, and a sensation commonly described as "bird legs," caused by the upward redistribution of bodily fluids [9]. Over the long term, microgravity can lead to changes in bone and muscle structure, with consequences for posture, balance, and physical strength. Astronauts may experience loss of bone and muscle mass, reduced bone density, weakening of postural muscles, and difficulty maintaining balance and coordination. Beyond the physical effects, microgravity can also influence body perception and self-image. Astronauts may feel a loss of control over their bodies and struggle to adapt to the new environmental conditions. Developing psychological interventions that support astronauts in accepting and adjusting to the physical changes caused by microgravity is essential. These strategies can encourage a more positive body image and enhance self-awareness of physical capabilities. Since adapting to microgravity demands considerable cognitive and physical effort, psychological support plays a key role in easing this transition [3].

2.1 Coping with stress and fatigue in space missions

Astronauts face elevated stress levels due to the demands and risks of space missions, the need to make rapid decisions in critical situations, and the constant pressure to meet objectives. Factors such as work overload, sleep deprivation, exposure to cosmic radiation, and the physical distance from Earth can lead to both physical and mental fatigue, diminishing performance, and increasing the likelihood of errors [10]. Chronic stress can negatively impact immune function, making astronauts more susceptible to illness, while also affecting their mental health by increasing the risk of anxiety, depression, and sleep disorders. Moreover, chronic stress can compromise cognitive abilities, impairing attention, memory, reasoning, and problem-solving skills [1]. Effective stress management is not only crucial for astronauts' well-being but also vital for the success of the mission. Communication delays with Earth, caused by the vast distance and current technological constraints, can lead to frustration, anxiety, a greater sense of isolation, and difficulties in solving complex problems. Without immediate support from mission control, astronauts face added challenges when making critical decisions [1]. In emergencies, these delays can have severe consequences, endangering astronaut safety and the success of the mission. Therefore, addressing the challenges posed by stress and communication delays requires a multifaceted approach. First, it is essential to incorporate stress-reducing techniques into the daily routines of astronauts. These can include mindfulness practices, regular physical exercise, and relaxation techniques that can help alleviate the physiological and psychological effects of stress [11]. Training in effective communication strategies and providing astronauts with tools for self-regulation, such as cognitive-behavioral methods, can also be critical in managing high-pressure situations. Moreover, preparing astronauts for the emotional and psychological strains of isolation is just as important as preparing them for technical and physical challenges. This preparation involves providing continuous psychological support, ensuring regular check-ins with mission control, and fostering a team environment that promotes open communication, mutual support, and conflict resolution. By focusing on these aspects, astronauts can maintain their mental health and well-being, thereby improving their ability to perform under pressure. Ultimately, the ability to manage stress and maintain a strong psychological state is not just an individual responsibility. It is a collective effort that involves mission planners, psychologists, and the astronauts themselves working together to create a support system that ensures the mental and physical well-being of the crew [12]. Such strategies are not just beneficial for the astronauts but are also essential for achieving the long-term success of space missions.

3. Psychological challenges of long-duration space missions

Beyond the challenges already discussed, it is essential to recognize that astronauts may encounter a range of additional psychological difficulties during long-duration space missions, each of which requires careful attention and targeted intervention strategies [13]. Among these, homesickness and longing for home emerge as powerful emotions, fueled by the distance from family, friends, and the comforting familiarity of everyday environments [5]. This prolonged separation can lead to deep sadness and an intense yearning to return to a safe, familiar place, negatively impacting mood and motivation [14]. Moreover, monotony and boredom can undermine astronauts' psychological well-being [15]. The repetitive nature of daily tasks, lack of environmental stimulation, and limited recreational opportunities may foster feelings of apathy and disengagement, reducing attention and focus [5]. Fear and uncertainty represent another significant challenge, driven by the inherent risks of space missions, uncertainty about the future, and the awareness of one's vulnerability in a hostile environment [16]. These emotions can increase anxiety and worry, negatively affecting decision-making and the ability to respond effectively to emergencies. Sleep difficulties, often caused by disrupted circadian rhythms, stress, noise, and lack of privacy, may impair rest quality and hurt mood, energy levels, and cognitive performance [17]. Lastly, interpersonal issues may arise due to the forced cohabitation in confined spaces, cultural differences, and communication challenges, potentially leading to conflict and misunderstandings within the crew. It is crucial to understand that these psychological challenges are not isolated phenomena; rather, they interact in complex ways, influencing overall well-being and performance [18, 19].

In addition, implementing structured and personalized daily routines can help combat monotony and boredom. These routines should incorporate physical exercise, which is known to improve mood, reduce stress, and enhance sleep, as well as recreational and creative activities that promote mental stimulation and a sense of purpose. Personalizing routines to reflect individual preferences can further increase motivation and engagement. Essential is the presence of robust social and psychological support systems, involving not only fellow crew members but also mission control personnel, family, and friends. A culture that encourages open communication and emotional expression can significantly reduce stress, anxiety, and feelings of isolation. Promoting crew interactions based on trust, empathy, and mutual support enhances cohesion and collective resilience. This approach should include a combination of psychological countermeasures aimed at promoting mental health, enhancing stress resilience, supporting cognitive performance, and strengthening group cohesion. Among the most promising strategies are mindfulness and relaxation techniques, which offer astronauts tools to manage stress and maintain emotional balance. Practices such as guided meditation, breathing exercises, progressive relaxation, and yoga have demonstrated benefits in improving stress resilience, emotional regulation, attention, and sleep quality [20]. Research found that regular mindfulness practice can reduce cortisol levels and activate brain regions involved in emotional regulation [21]. The MINDFUL-ICE study, conducted in an isolated and confined analog environment, further reinforces the potential of mindfulness-based strategies to mitigate stress during long-term missions [2]. Professional psychological support from specialists trained in space psychology should also be made accessible, ensuring that astronauts have access to expert care when needed. Feeling connected to a larger support network and having the opportunity to share concerns can foster a sense of belonging and psychological safety. The MINDFUL-ICE project also confirms the importance of social support in mitigating overall stress and preserving emotional well-being. Therefore, a holistic approach is needed, one that considers all contributing factors to stress and difficulty during space missions. This approach should include personalized interventions tailored to individual needs and promote a supportive, collaborative environment among crew members.

3.1 Keeping balance: The role of routine in space missions

In the demanding context of long-duration space missions, where environmental stressors and limited resources can challenge cognitive and emotional stability, implementing structured daily routines emerges as one of the critical countermeasures. These routines, thoughtfully incorporating regular exercise, VR environments for mood enhancement, and creative leisure activities, anchor normalcy and predictability, fostering a sense of control and contributing to overall mental balance. Regular physical exercise has well-established benefits for mood regulation and stress reduction [22, 23] and can promote psychological well-being. Furthermore, research indicates that exercise can mitigate the negative impact of isolation and confinement on cognitive function. The scheduled integration of VR environments, simulating natural landscapes or engaging social interactions, offers astronauts a valuable escape from the confines of the spacecraft and can alleviate feelings of loneliness and boredom. Creative leisure activities, such as painting, writing, playing musical instruments, or engaging in virtual collaborative projects with ground-based teams, provide opportunities for self-expression, cognitive stimulation, and a sense of accomplishment. These activities not only combat monotony but also foster resilience by promoting a sense of purpose and connection to one's identity beyond the mission. Subject matter experts considered mindfulness as a particularly relevant construct for space exploration [11]. This awareness of self can be further bolstered with structured routines. The effectiveness of structured routines lies in their capacity to provide a sense of predictability and control in an environment characterized by uncertainty and constraint. These habits provide a framework, and by providing a schedule, the astronaut does not need to stress time management. Furthermore, these routines promote self-care and encourage astronauts to actively engage in activities that enhance their well-being, thereby mitigating the negative psychological consequences of long-duration spaceflight. A recent systematic review [24] on analog environments has emphasized the need for additional analog studies, particularly in longer duration contexts and in extreme condition contexts.

3.2 The role of vulnerability in elite performance

It is essential to emphasize that, despite rigorous selection and extensive training, astronauts remain human and inherently vulnerable to psychological and physiological stressors. The high expectations placed on space crews by stakeholders often obscure the reality of mental health challenges associated with isolation, confinement, and mission pressure. Expecting continuous high performance without robust psychological support is unrealistic. Recognizing the limits of human resilience and providing structured, ongoing psychological care are fundamental to safeguarding astronaut well-being and mission success.

The perception of astronauts as supremely resilient individuals capable of enduring extreme conditions can inadvertently mask the potential for mental health challenges. Unlike machines, human beings are not designed for constant, unwavering performance. Even with the most advanced training, astronauts remain susceptible to stress, anxiety, depression, and other psychological difficulties exacerbated by the unique stressors of spaceflight [5, 14]. This inherent vulnerability necessitates a shift in perspective, moving away from the expectation of superhuman resilience toward an acknowledgment of the human element in space exploration. For instance, studies have shown that even short-duration space missions can lead to significant changes in brain structure and function, impacting mood, cognitive performance, and emotional regulation [5]. During long-duration missions, these effects are amplified, potentially leading to chronic stress, burnout, and impaired decision-making. Recognizing these potential impacts is crucial for mission planners, healthcare professionals, and astronauts themselves to ensure adequate psychological support is available throughout the mission.

Furthermore, the pressure to maintain a flawless public image and to downplay any signs of vulnerability can further exacerbate the psychological toll on astronauts. Encouraging open communication, promoting psychological self-care, and creating a supportive team environment are essential for fostering a culture where astronauts feel safe seeking help without fear of judgment or professional repercussions [25]. The goal is not to diminish the achievements or capabilities of astronauts but to acknowledge their humanity and to provide the necessary resources to sustain their mental health and well-being in the face of extraordinary challenges.

4. Managing stress in space: The role of mindfulness and relaxation techniques

As highlighted, long-duration space missions expose astronauts to a unique set of psychological stressors, including isolation, confinement, circadian rhythm disruptions, and the monotony of a highly controlled environment. These conditions can lead to elevated levels of stress hormones, impaired immune function, and a decline in cognitive performance, all of which threaten mission success and crew well-being [26]. In recent years, mindfulness and relaxation techniques have emerged as promising, evidence-based strategies to mitigate these risks and promote psychological resilience during spaceflight. Mindfulness, defined as the practice of maintaining a non-judgmental awareness of the present moment, has been extensively studied in terrestrial settings for its capacity to reduce stress, improve emotional regulation, and enhance cognitive flexibility [24]. Its application to spaceflight is grounded in both empirical research and analog studies conducted in environments such as Antarctica's Concordia base, where prolonged isolation and environmental extremes closely mimic those of space missions [2]. In these settings, individuals with a higher disposition toward mindfulness have demonstrated greater stress resistance and improved adaptation to challenging conditions [24]. The physiological and psychological benefits of mindfulness are multifaceted. Regular practice has been shown to attenuate the stress response by reducing glucocorticoid secretion, which in turn helps protect immune function and supports neurological integrity [26]. Neuroimaging studies suggest that mindfulness meditation can induce structural and functional changes in brain regions associated with attention, emotional regulation, and executive function, even after short training periods [26]. For astronauts, these adaptations may translate into improved focus, better decision-making under pressure, and enhanced capacity to manage the emotional challenges inherent to space missions [27]. Relaxation techniques such as diaphragmatic breathing and progressive muscle relaxation complement mindfulness by directly targeting the physiological manifestations of stress. Slow, deep breathing activates the parasympathetic nervous system, reducing heart rate and blood pressure, while progressive muscle relaxation systematically releases tension throughout the body [1]. These methods are particularly well-suited to the space environment: they are noninvasive, require no special equipment, and can be practiced in confined quarters with minimal time commitment [1]. Importantly, these techniques have also been linked to improved sleep quality, which is often disrupted in space due to environmental factors and irregular work schedules [1]. The integration of mindfulness and relaxation training into astronaut preparation is increasingly recognized as a critical component of crew health support. Space agencies now recommend that astronauts begin mindfulness practice several months before launch to maximize neurobehavioral benefits and continue these routines during missions. Protocols typically involve brief, regular sessions, such as 5–10 minutes of mindfulness meditation or breathing exercises every 8 hours, tailored to the operational constraints of spaceflight [28, 29]. Some programs also explore the use of immersive technologies and biofeedback to enhance engagement and monitor effectiveness in real time [24]. Despite logistical challenges, such as time constraints and the need for adaptation to microgravity and spacecraft noise, the evidence supports the feasibility and utility of these interventions in space. As missions grow longer and venture farther from Earth, these practices will play an increasingly central role in sustaining astronaut health, optimizing cognitive function, and ensuring the overall success of human spaceflight [1, 24].

4.1 Applying mindfulness research from Antarctica to spaceflight

Analog environments, which replicate key stressors of space missions, play a critical role in developing and testing effective psychological countermeasures. Antarctica's Concordia research station, often called "White Mars," provides a high-fidelity analog due to its extreme isolation, confinement, and harsh environmental conditions [30]. A recent study [2] leveraged this unique setting to investigate the link between mindfulness and stress among Concordia crew members. The study assessed 24 individuals over a 12-month mission using multiple measures of mindfulness and perceived stress. Results showed that those with a higher disposition toward mindfulness reported significantly lower levels of perceived stress. The findings revealed a strong negative association between mindfulness and stress, suggesting that mindfulness may play a crucial role in alleviating stress in isolated and confined environments, such as those encountered during long-duration space missions. By highlighting the potential for early identification of resilient individuals, the study supports the integration of mindfulness assessments into astronaut selection procedures to identify candidates best equipped to handle the psychological demands of extended spaceflight. These results suggest that fostering mindfulness could be an effective strategy for reducing the psychological challenges faced by astronauts during long-duration missions. Moreover, the study highlights the potential for incorporating mindfulness assessments into astronaut selection procedures, identifying candidates who are naturally more resilient to stress in isolated and confined environments. Applying findings from Antarctic research to the context of spaceflight can help enhance astronauts' preparedness for the psychological and emotional challenges of long-duration missions.

4.2 Practical techniques for building resilience and managing stress in space

Alongside the broader benefits of a mindful disposition, specific techniques offer practical tools for managing stress in the demanding environment of space. Relaxation techniques, such as diaphragmatic breathing and progressive muscle relaxation, provide astronauts with readily accessible methods for directly addressing the physiological manifestations of stress [1]. Slow, deep breathing activates the parasympathetic nervous system, promoting relaxation and reducing heart rate and blood pressure. Progressive muscle relaxation systematically releases tension throughout the body, easing physical discomfort and promoting a sense of calm. These techniques are particularly well-suited for space missions due to their noninvasive nature, minimal equipment requirements, and adaptability to confined spaces [1]. Regular practice of these techniques has also been linked to improved sleep quality, which is often compromised in space due to disrupted circadian rhythms and demanding work schedules. Integrating mindfulness and relaxation training into astronaut preparation allows space agencies to equip crews with self-regulation tools that enhance resilience and support optimal performance during long-duration missions.

5. New ways to support astronauts' psychological well-being

While the concept of providing psychological support during space missions is well-established, the unique challenges of deep space exploration demand a more sophisticated and adaptive approach [31]. The limitations of real-time communication with Earth-based psychologists due to significant time delays necessitate a shift toward empowering astronauts with self-sufficiency and proactive coping mechanisms. Future strategies must encompass a multi-tiered system that integrates personalized interventions, robust peer support networks, and cutting-edge technological solutions, all while adhering to strict ethical guidelines [32]. Individual responses to the space environment vary significantly, highlighting the need for tailored support. Future research should aim to develop flexible systems that integrate real-time biofeedback, such as heart rate variability, skin conductance, or breathing patterns, with the analysis of communication dynamics. These systems could help detect individual sources of stress, whether emotional, cognitive, or situational, and respond in a timely and personalized way. Combining physiological monitoring with the study of interaction patterns would allow for more accurate and context-sensitive interventions, with potential applications in settings where well-being and performance are closely linked, such as education, healthcare, or workplace environments. Furthermore, it is essential to consider cultural differences, as individuals from diverse backgrounds may rely on different strategies to cope with stress and may express their needs for support in distinct ways. Recognizing and respecting these variations can contribute to creating a more inclusive and responsive environment, where each person feels understood and supported according to their cultural frame of reference [33]. This becomes particularly relevant in international and high-stress contexts, such as long-duration space missions, where effective communication and psychological well-being are key. Robust peer support networks can serve as a valuable complement to professional psychological services. Training astronauts in active listening, conflict resolution, and psychological first aid can foster a more resilient and cohesive team environment [1]. However, it is essential to acknowledge the potential challenges of peer support, such as boundary violations and vicarious traumatization, and implement safeguards to mitigate these risks [34]. Guidelines for peer support should be developed and integrated into pre-flight training. The integration of technology, such as Artificial Intelligence (AI)-driven therapy systems, holds an immense promise for providing on-demand mental health support [11]. These systems could offer personalized cognitive-behavioral therapy (CBT), mindfulness exercises, and stress-reduction techniques tailored to individual needs. However, the ethical implications of using such technologies must be carefully considered. Data privacy, algorithmic bias, and the potential misuse of psychological information are critical concerns that require proactive attention. Ethical and responsible implementation of psychological interventions is essential. Furthermore, it suggests that mindfulness assessment could be added to crew selection procedures [2]. Finally, ongoing research is needed to evaluate the long-term effects of spaceflight on psychological well-being, since longitudinal studies that track astronauts' mental health for years after their missions can provide valuable insights into the potential long-term consequences of space travel and inform the development of more effective countermeasures.

5.1 The role of virtual reality in psychological support

VR technology offers unique possibilities for enhancing psychological support during space missions. VR environments can be used to simulate Earth-like settings, providing astronauts with a sense of connection to home and reducing feelings of isolation [35]. VR can also be used for therapeutic interventions, such as exposure therapy for anxiety or stress-reduction exercises in immersive, calm environments. Moreover, VR can facilitate social interaction by allowing astronauts to connect with family and friends in a virtual space, creating a sense of presence and reducing feelings of loneliness. One such VR program helped astronauts cope with interpersonal conflict and depression while engaged in deep space missions [29, 36]. However, the use of VR in space also presents challenges, such as motion sickness, technical limitations, and the need for careful design to ensure that VR experiences are engaging and beneficial. NASA has been proactively testing the usability of VR technologies in these types of settings [37].

5.1.1 Automated psychotherapy in space and its impact on mental healthcare

As human space exploration extends beyond low Earth orbit, the challenges of providing adequate mental healthcare to astronauts grow more complex. The substantial communication delays inherent in deep space missions make traditional Earth-based psychotherapy less feasible, necessitating innovative approaches that can deliver timely and effective support [1]. Automated psychotherapy systems offer a promising solution, providing astronauts with immediate, confidential access to evidence-based interventions tailored to the unique stressors of space travel. Recognizing the need for accessible mental health support during long-duration space missions, several early initiatives have focused on developing and testing automated psychotherapy tools. These projects aim to offer astronauts personalized interventions to monitor their behavior, performance, and emotional state, suggesting routine adjustments to mitigate stress and reduce the risk of developing psychopathologies [38]. For example, interactive media programs are being developed to help astronauts manage interpersonal conflict and depression [39]. One such program, tested in a space analog environment with a crew of six individuals isolated for 8 months, demonstrated the effectiveness of the conflict and stress modules. Participants expressed a strong desire to use these tools to support their fellow crew members. In addition to interactive programs, virtual and augmented reality (VR/AR) methods are being explored to address psychological challenges. The Crew Interactive Mobile Companion (CIMON), an AI assistant developed by IBM and Airbus Group, has been tested on the International Space Station. While primarily designed to offer task guidance and answer technical questions [28], CIMON's use in space is an early example of integrating technology to support astronauts' mental well-being. Despite these promising developments, the field of automated psychotherapy in space remains in its infancy. Limited publicly available information regarding the specific tools currently used by astronauts during missions highlights the need for greater transparency from NASA and other space agencies. If such tools are not yet in use during actual spaceflight, it is crucial to prioritize their integration as soon as possible, drawing on insights from analog environment studies like the MINDFUL-ICE project at Concordia Station [2]. This project underscores the importance of mindfulness in mitigating stress in isolated and confined environments. One of the primary advantages of automated psychotherapy systems is their ability to overcome the limitations of communication delays, ensuring that astronauts can receive support whenever and wherever it is needed [28]. These systems can provide a variety of benefits, including on-demand CBT techniques, stress management strategies, and guided mindfulness exercises. Additionally, they can be designed to adapt to individual astronaut needs and preferences, offering personalized interventions based on real-time data and feedback. Automated systems can also be integrated with other mental health resources, such as wearable sensors that monitor physiological indicators of stress and VR environments that simulate supportive social interactions [1, 3]. However, the development and deployment of automated psychotherapy systems in space also face significant challenges. Ensuring the efficacy and safety of these systems requires rigorous testing and validation in analog environments that replicate the stressors of spaceflight. Ethical concerns, including data privacy, algorithmic bias, and the potential for over-reliance on technology, must also be carefully addressed. Despite these challenges, automated psychotherapy holds great promise for bridging the gap in mental healthcare for astronauts, as promoting psychological well-being through these systems can help ensure the success of long-duration space missions [1].

6. The role of crew selection and training in promoting psychological resilience

While technological solutions and on-demand support systems are essential, proactive measures taken before a mission can significantly enhance the psychological resilience of astronauts. The crew selection process offers a crucial opportunity to identify individuals with pre-existing strengths, such as a disposition toward mindfulness [1]. Integrating mindfulness assessments into the selection process, as suggested by the MINDFUL-ICE study, could help identify candidates who are naturally better equipped to manage the stressors of isolation and confinement [2]. These programs should not only educate astronauts about potential psychological challenges but also provide them with practical tools and techniques to proactively address these challenges during the mission. Analog environments like Concordia Station provide an opportunity for researchers to conduct these types of training before embarking into space [24]. Investing in robust crew selection and training allows space agencies to establish a strong foundation for promoting psychological well-being throughout a mission.

7. The evolving landscape of psychological support in space

While the core elements of psychological support are well-established, several subtle yet crucial factors warrant deeper exploration. These considerations become increasingly important as mission durations extend and crews venture further from Earth, increasing the challenges associated with traditional support models. One key area of focus is the personalization of psychological interventions [1]. While automated systems and pre-flight training can provide a baseline level of support, they may not adequately address the unique needs and experiences of individual astronauts. Factors such as personality traits, coping styles, and cultural background can significantly influence how astronauts respond to the stressors of spaceflight. Therefore, future research should prioritize the development of adaptive systems that can tailor interventions to meet the specific needs of each crew member. This could involve incorporating biofeedback sensors to monitor physiological responses to stress or using machine learning algorithms to analyze communication patterns and identify potential psychological distress. Another crucial aspect is the integration of peer support programs. Astronauts often form close bonds with their crewmates, and these relationships can serve as a valuable source of emotional support during long-duration missions [18]. To enhance the impact of peer support, it is crucial to provide astronauts with targeted training that develops key interpersonal and emotional competencies. This includes learning how to listen empathetically, manage conflicts constructively, and identify early signs of psychological strain within the team. Cultivating these skills not only strengthens mutual trust but also contributes to a more resilient and supportive crew dynamic, helping to buffer the psychological toll of isolation and confinement. Furthermore, the ethical implications of using advanced technologies for psychological support in space require careful consideration. Issues such as data privacy, algorithmic bias, and the potential for misuse of psychological information must be addressed proactively to ensure that these technologies are used responsibly and ethically. Transparency, accountability, and independent oversight are essential to maintaining the trust of astronauts and the public. Ongoing research is crucial to fully understand the long-term psychological effects of spaceflight. Longitudinal studies that follow astronauts' mental health well beyond their missions can offer valuable insights into the lasting impact of life in space and help shape more effective support strategies. By committing to research, developing tailored interventions, strengthening peer support initiatives, and establishing clear ethical guidelines, it is possible to build a comprehensive and enduring psychological support system—one that safeguards astronauts' mental health and helps them adapt and thrive in the demanding conditions of space.

8. Terrestrial applications of space psychology

Many psychological countermeasures developed to support astronauts-such as mindfulness, relaxation techniques, and automated mental health systems-have significant value on Earth. These approaches are increasingly applied in high-stress or isolated environments, including healthcare, education, remote work, and polar research stations. The transfer of knowledge from space psychology to terrestrial contexts highlights the broader societal relevance of investing in astronaut psychological health offering effective tools to enhance well-being and resilience in diverse populations. The healthcare sector, characterized by high-pressure environments and emotionally demanding situations, has increasingly adopted psychological interventions initially designed for astronauts. For instance, mindfulness-based stress reduction (MBSR) programs, which have proven effective in reducing stress and enhancing cognitive function in spaceflight simulations [40], are now widely used to combat burnout among nurses and physicians [41]. Similarly, virtual reality therapies, initially developed to mitigate the effects of isolation and confinement during space missions, are now used to manage pain, anxiety, and post-traumatic stress disorder (PTSD) in clinical settings [42]. The educational field also benefits from insights gained in space psychology. Strategies for enhancing focus, reducing stress, and promoting teamwork, essential for astronauts in confined environments, are now applied in classrooms and online learning platforms. Techniques such as guided meditation and biofeedback, which help astronauts regulate their emotional states [42], are used to improve attention and reduce anxiety among students [43]. Moreover, collaborative problem-solving activities, modeled after crew resource management training in space missions, are implemented to foster communication and cooperation among students [2]. The rise of remote work has created new challenges related to isolation, stress, and maintaining work-life balance Space psychology offers valuable strategies for addressing these issues. For example, structured daily routines, a key element in astronaut well-being [2], are now promoted to remote workers to enhance productivity and reduce feelings of isolation. Similarly, virtual social support systems, which provide astronauts with a sense of connection to Earth, are replicated in online communities and virtual co-working spaces to foster social interaction among remote employees [44]. Polar research stations, characterized by extreme isolation and harsh environmental conditions, share many similarities with space missions. Psychological interventions developed for astronauts, such as stress management techniques and team-building exercises, are directly applicable in these settings. Studies have shown that mindfulness training can improve mood and reduce stress among researchers in Antarctica [24]. Furthermore, strategies for enhancing crew cohesion and conflict resolution, crucial for spaceflight teams, are implemented to foster positive interpersonal relationships among polar researchers [45]. The translation of psychological knowledge and countermeasures from space exploration to terrestrial settings underscores the broader societal impact of investing in astronaut health and well-being. Adapting and implementing these strategies across various contexts can enhance resilience, promote mental health, and improve performance in populations facing high-stress or isolated conditions.

9. Conclusion

The psychological needs of astronauts are a crucial factor to consider for the success of future space missions. Implementing effective psychological countermeasures can enhance astronauts' well-being, increase stress resilience, ensure optimal cognitive performance, foster crew cohesion, and help maintain a positive and collaborative environment. Further research is needed to develop and evaluate new psychological support strategies, considering the specific challenges of long-duration missions to the Moon and Mars. It is important to investigate the effectiveness of emerging technologies, such as artificial intelligence, virtual reality, and wearable devices, in providing psychological support to astronauts. Moreover, psychological countermeasures must be tailored to individual needs and adapted to the unique characteristics of each mission. Only through an integrated and personalized approach will it be possible to ensure the psychological well-being and mission success of future generations of space explorers. It is essential to invest in research and the development of innovative support strategies to prepare astronauts for the psychological demands of long-duration missions and to enable them to have a positive and meaningful experience. The future of space exploration depends not only on technological advancements but also on our ability to protect and promote astronauts' psychological health.

Acknowledgements

This work was supported by the Italian Space Agency, through the projects RelaxPro (DC-VUM-2020-007, Mind/body practices for deep space exploration) and MINDFUL-ICE II (DC-DSR-UVS-2022-212, "A Quick Mindfulness Training for an Isolated and Confined Environment"), the European Space Agency, the French Polar Institute Paul-Émile Victor, and the Programma Nazionale di Ricerche in Antartide.

Conflict of interest

The authors declare no conflict of interest.

Author details

Francesco Pagnini* and Francesca Grosso
Department of Psychology, Università Cattolica del Sacro Cuore, Milan, Italy

*Address all correspondence to: francesco.pagnini@unicatt.it

References

[245] Pagnini F. Supporting the mind in space: Psychological tools for long-duration missions. Interactive Journal of Medical Research. 2024;**13**:e66626. DOI: 10.2196/66626

[246] Pagnini F, Thoolen S, Smith N, Van Ombergen A, Grosso F, Langer E, et al. Mindfulness disposition as a protective factor against stress in Antarctica: A potential countermeasure for long-duration spaceflight? Journal of Environmental Psychology. 2024;**94**:102254. DOI: 10.1016/J.JENVP.2024.102254

[247] Pagnini F, Manzey D, Rosnet E, Ferravante D, White O, Smith N. Human behavior and performance in deep space exploration: Next challenges and research gaps. npj Microgravity. 2023;**9**:1-7. DOI: 10.1038/s41526-023-00270-7

[248] Blue RS, Jennings RT, Antunano MJ, Mathers CH. Commercial spaceflight: Progress and challenges in expanding human access to space. REACH. 2017;**7-8**:6-13. DOI: 10.1016/j.reach.2018.08.001

[249] Arone A, Ivaldi T, Loganovsky K, Palermo S, Parra E, Flamini W, et al. The burden of space exploration on the mental health of astronauts: A narrative review. Clinical Neuropsychiatry. 2021;**18**(5):237-246. DOI: 10.36131/cnfioritieditore20210502

[250] Palinkas L, Suedfeld P. Psychological effects of polar expeditions. Lancet. 2008;**371**:153-163. DOI: 10.1016/S0140-6736(07)61056-3

[251] Marcinkowski M, Bell S, Roma P. The nature of conflict for teams in isolated, confined, and extreme environments. Acta Astronautica. 2021;**181**:81-91. DOI: 10.1016/j.actaastro.2021.01.004

[252] Tays GD, Hupfeld KE, McGregor HR, Banker LA, De Dios YE, Bloomberg JJ, et al. The microgravity environment affects sensorimotor adaptation and its neural correlates. Cerebral Cortex. 2025;**35**:bhae502. DOI: 10.1093/cercor/bhae502

[253] Clément GR, Bukley AP, Paloski WH. Artificial gravity as a countermeasure for mitigating physiological deconditioning during long-duration space missions. Frontiers in Systems Neuroscience. 2015;**9**:1-11. DOI: 10.3389/fnsys.2015.00092

[254] Jimenez MP, Deville NV, Elliott EG, Schiff JE, Wilt GE, Hart JE, et al. Associations between nature exposure and health: A review of the evidence. International Journal of Environmental Research and Public Health. 2021;**18**:2-19. DOI: 10.3390/ijerph18094790

[255] Pagnini F, Phillips D, Katherine B, Ellen L. Mindfulness and relaxation training for long duration spaceflight: Evidences from analog environments and military settings. Acta Astronautica. 2019;**165**:1-8. DOI: 10.1016/j.actaastro.2019.07.036

[256] Landon LB, Slack KJ, Barrett JD. Teamwork and collaboration in long-duration space missions: Going to extremes. American Psychologist. 2018;**73**:563-575. DOI: 10.1037/amp0000260

[257] Basner M, Babisch W, Davis A, Brink M, Clark C, Janssen S, et al. Auditory and non-auditory effects of noise on health. The Lancet. 2014;**383**:1325-1332. DOI: 10.1016/S0140-6736(13)61613-X

[258] Yin Y, Liu J, Fan Q, Zhao S, Wu X, Wang J, et al. Long-term spaceflight composite stress induces depression and cognitive impairment in astronauts—Insights from neuroplasticity. Translational Psychiatry. 2023;**13**:342. DOI: 10.1038/s41398-023-02638-5

[259] Tomsia M, Cieśla J, Śmieszek J, Florek S, Macionga A, Michalczyk K, et al. Long-term space missions' effects on the human organism: What we do know and what requires further research. Frontiers in Physiology. 2024;**15**:1-19. DOI: 10.3389/fphys.2024.1284644

[260] Sandal G, Van de Vijver F, Smith N. Psychological hibernation in antarctica. Frontiers in Psychology. 2018;**9**:1-8. DOI: 10.3389/fpsyg.2018.02235

[261] Albornoz-Miranda M, Parrao D, Taverne M. Sleep disruption, use of sleep-promoting medication and circadian desynchronization in spaceflight crewmembers: Evidence in low-Earth orbit and concerns for future deep-space exploration missions. Sleep Medicine: X. 2023;**6**:1-5. DOI: 10.1016/j.sleepx.2023.100080

[262] Heinicke C, Poulet L, Dunn J, Meier A. Crew self-organization and group-living habits during three autonomous, long-duration Mars analog missions. Acta Astronautica. 2021;**182**:160-178. DOI: 10.1016/j.actaastro.2021.01.049

[263] Kanas N, Salnitskiy V, Grund E, Gushin V, Weiss D, Kozerenko O, et al. Interpersonal and cultural issues involving crews and ground personnel during Shuttle/Mir space missions. Aviation, Space, and Environmental Medicine. 2000;**71**:A11-A16

[264] Grossman P, Tiefenthaler-Gilmer U, Raysz A, Kesper U. Mindfulness training as an intervention for fibromyalgia: Evidence of postintervention and 3-year follow-up benefits in well-being. Psychotherapy and Psychosomatics. 2007;**76**:226-233. DOI: 10.1159/000101501

[265] Pagnini F, Philips D. Being mindful about mindfulness. Lancet Psychiatry. 2015;**2**(4):288-289

[266] De la Torre GG, Groemer G, Diaz Artiles A, Pattyn N, Van Cutsem J, Musilova M, et al. Space analogs and behavioral health performance research review and recommendations checklist from ESA Topical Team. npj Microgravity. 2024;**10**:98. DOI: 10.1038/s41526-024-00437-w

[267] Crucian BE, Choukèr A, Simpson RJ, Mehta S, Marshall G, Smith SM, et al. Immune system dysregulation during spaceflight: Potential countermeasures for deep space exploration missions. Frontiers in Immunology. 2018;**9**:1-21. DOI: 10.3389/fimmu.2018.01437

[268] Le Roy B, Martin-Krumm C, Trousselard M. Mindfulness for adaptation to analog and new technologies emergence for long-term space missions. Frontiers in Space Technologies. 2023;**4**:1-11. DOI: 10.3389/frspt.2023.1109556

[269] Ann RK. Astronauts' sensemaking of dangerous beauty: An account of the overview effect for organizational theory. The Journal of Applied Behavioral Science. 2022;**60**:494-519. DOI: 10.1177/00218863221136289

[270] Saniotis A, De la Torre GG, Galassi F, Henneberg M, Mohammadi K. Mindfulness meditation and spaceflight: A potential adjunct therapy for astronauts. THE MIND - Bulletin on Mind-Body Medicine Research. 2023;**3**: 1-4. DOI: 10.61936/themind/202312222

[271] Pendergraft J, Carter D, Tseng S, Landon L, Slack K, Shuffler M. Learning from the past to advance the future: The adaptation and resilience of NASA's spaceflight multiteam systems across four eras of spaceflight. Frontiers in Psychology. 2019;**10**:1-23. DOI: 10.3389/fpsyg.2019.01633

[272] Smith L. Space station and spacecraft environmental conditions and human mental health: Specific recommendations and guidelines. Life Sciences in Space Research. 2024;**40**:126-134. DOI: 10.1016/j.lssr.2023.10.001

[273] Ablon JS, Levy RA, Smith-Hansen L. The Contributions of the Psychotherapy Process Q-set to Psychotherapy Research. Research in Psychotherapy. 2011;**14**:14-48

[274] Van Ombergen A, Rossiter A, Ngo-Anh TJ. 'White Mars'—Nearly two decades of biomedical research at the Antarctic Concordia station. Experimental Physiology. 2021;**106**:6-17. DOI: 10.1113/EP088352

[275] Manzey D. Human missions to Mars: New psychological challenges and research issues. Acta Astronautica. 2004;**55**:781-790. DOI: 10.1016/j.actaastro.2004.05.013

[276] Kanas N. Psychological, psychiatric, and interpersonal aspects of long-duration space missions. Journal of Spacecraft and Rockets. 1990;**27**:457-463. DOI: 10.2514/3.26165

[277] Ritsher JB. Cultural factors and the International Space Station. Aerospace Medicine and Human Performance. Jun 2005;**76**(6 Suppl):B135-44

[278] Dobney W, Mols L, Mistry D, Tabury K, Baselet B, Baatout S. Evaluation of deep space exploration risks and mitigations against radiation and microgravity. Frontiers in Nuclear Medicine. 2023;**3**:1-21. DOI: 10.3389/fnume.2023.1225034

[279] Sharp J, Kelson J, South D, Saliba A, Kabir MA. Virtual reality and artificial intelligence as psychological countermeasures in space and other isolated and confined environments: A scoping review. Acta Astronautica. 2025;**232**:666-677. DOI: 10.1016/j.actaastro.2025.04.002

[280] Pagnini F. Psychological wellbeing and quality of life in amyotrophic lateral sclerosis: A review. International Journal of Psychology. 2013;**48**:194-205. DOI: 10.1080/00207594.2012.691977

[281] Brent Woodland M, Ong J, Zaman N, Hirzallah M, Waisberg E, Masalkhi M, et al. Applications of extended reality in spaceflight for human health and performance. Acta Astronautica. 2024;**214**:748-756. DOI: 10.1016/j.actaastro.2023.11.025

[282] Schmidt MA, Goodwin TJ. Personalized medicine in human space flight: Using omics based analyses to develop individualized countermeasures that enhance astronaut safety and performance. Metabolomics. 2013;**9**:1134-1156. DOI: 10.1007/s11306-013-0556-3

[283] Porte Y, Morel JL. Learning on Jupiter, learning on the moon: The dark side of the G-force. Effects of gravity changes on neurovascular unit and modulation of learning and memory. Frontiers in Behavioral Neuroscience. 2012;**6**(64):1-6. DOI: 10.3389/fnbeh.2012.00064

[284] Strangman GE, Sipes W, Beven G. Human cognitive performance in spaceflight and analogue environments. Aerospace Medicine and Human Performance.

2014;**85**(10):1033-1048. DOI: 10.3357/ASEM.3961.2014

[285] Shapiro SL, Schwartz GE, Bonner G. Effects of mindfulness-based stress reduction on medical and premedical students. Journal of Behavioral Medicine. 1998;**21**:581-599. DOI: 10.1023/A:1018700829825

[286] Rotondo G, Pancheri P, Monesi F, Grantaliano G, Depascalis V. Biofeedback monitoring-devices for astronauts in space environment. Acta Astronautica. 1983;**10**(8):591-598

[287] Lidia Z, Ackerman Deborah L, Yang May H, Futrell Julie L, Horton Nancy L, Sigi HT, et al. Mindfulness meditation training in adults and adolescents with ADHD: A feasibility study. Journal of Attention Disorders. 2007;**11**:737-746. DOI: 10.1177/1087054707308502

[288] Piko B, Kiss H, Hartmann A, Hamvai C. The role of social comparison and online social support in social media addiction mediated by self-esteem and loneliness. European Journal of Mental Health. 2024;**19**:1-11. DOI: 10.5708/EJMH.19.2024.0019

[289] Gangeme A, Simpson B, De la Torre GG, Larose T, Diaz Artiles A. A comprehensive look behind team composition for long duration spaceflight. Aerospace Medicine and Human Performance. 2023;**94**:457-465. DOI: 10.3357/AMHP.6175.2023

Chapter 8

Resilient by Design: Mitigating Human-Centered Risks in Space Missions and Critical Infrastructure

Ulpia-Elena Botezatu and Victor Vevera

Abstract

Long-duration space missions pose unique human-centered risks across cognitive, psychological, behavioral, and operational domains. This chapter reviews these vulnerabilities—including the effects of isolation, confinement, stress, and autonomy demands—and examines how design and training can bolster human resilience. We synthesize the empirical findings from International Space Station (ISS) expeditions, space analog simulations (e.g., Mars-500, HI-SEAS, Concordia), and Artemis mission planning. Space agencies emphasize that as missions move beyond low Earth orbit, crews face extended durations, communication delays, and increased autonomy requirements. Cognitive domains such as processing speed, working memory, attention, and decision-making can be subtly affected, though high-performing astronauts on six-month ISS missions generally maintain overall cognitive stability. Nonetheless, spaceflight stressors (environmental and mission challenges such as elevated CO_2 levels, sleep disruption, radiation, high workload) can degrade mood and cognitive function. Socially, small crews must manage conflicts, cultural differences, and limited privacy. Human factors experts argue that designing systems with human needs in mind (human–systems integration) and expecting inevitable human error are critical: Systematic human-centered design, combined with error-tolerant processes, improves resilience. We identify the gaps in current mitigation strategies—such as insufficient training for team cohesion and autonomy—and propose the recommendations: adaptive pre- and in-flight training, enhanced selection for behavioral resilience, autonomous mental health support tools, and design of habitats and procedures that maximize the crew well-being. By placing humans at the center of system design and adopting proactive resilience strategies, future space missions can achieve safer, more adaptable operations beyond Earth orbit.

Keywords: space systems resilience, critical infrastructure protection, human missions, space risk management, cyber-physical threats, space sustainability

1. Introduction

Human space exploration in the twenty-first century is entering a new era of distance and duration. NASA's Artemis program and planned crewed missions to Mars will send astronauts farther from Earth and for longer periods than ever before. With this expansion comes amplified human risks: As missions extend beyond low Earth orbit, astronauts will endure prolonged isolation, confinement, and communication delays, with no possibility of rapid evacuation [1, 2]. The NASA Human System Risk Board notes that risks scale with mission scope—what was manageable on the ISS (weeks to months in low Earth orbit) becomes more severe during year-long lunar or multi-year Mars missions [1]. Extended weightlessness and radiation exposure also add physiological burdens that interact with the psychological stressors.

Indeed, human factors in spaceflight constitute a highly specialized subfield, distinct from terrestrial human factors. Conditions such as microgravity (prolonged weightlessness), cosmic radiation, extreme isolation, and multi-minute communication delays have no parallel on Earth [3]. These factors necessitate space-specific solutions: For example, spacecraft interfaces must be operable in weightlessness (often by astronauts in bulky gloves), safety protocols must account for the impossibility of immediate evacuation or real-time ground support, and all systems and procedures require validation under space-like conditions to ensure reliability far from Earth. Designing for astronaut performance and safety thus demands tailored approaches beyond those of conventional workplaces.

Success in deep space hinges on the crew's ability to adapt and perform in these extreme conditions. Humans must execute complex tasks under high stakes, where even small errors can be catastrophic [4]. Thus, understanding cognitive vulnerabilities (e.g., memory, attention, decision-making) and behavioral factors (e.g., stress, group dynamics, autonomy) is essential. Equally, space systems must be designed for human resilience. This chapter reviews the state of knowledge on human-centered risks in spaceflight, drawing on space missions (ISS, Mir), ground analogues (Antarctic stations, underwater habitats, isolation chambers), and agency reports. We examine the interplay of psychological stress, team functioning, human–automation interaction, and mission operations. Finally, we identify gaps in mitigation (e.g., limited research on extremely long missions, insufficient training for autonomy) and offer recommendations for training, mission planning, and policy to bolster psychological resilience and safety. Additionally, we introduce how these spaceflight lessons can inform resilience in critical infrastructure on Earth, framing the final section as a forward-looking perspective rather than a separate topic.

2. Key concepts and definitions

Stressors in the spaceflight context are environmental, operational, and psychological factors that challenge astronaut health, performance, and well-being. These include microgravity, elevated CO_2, radiation exposure, noise, sleep disruption, isolation, and high workload. Unlike ordinary workplace stressors, these conditions are often chronic, cumulative, and unavoidable [5].

Stresses refer to the physiological and psychological responses that astronauts experience when exposed to stressors. These can manifest as fatigue, mood changes, irritability, or decreased cognitive performance. Stresses are not inherently negative—they may trigger adaptation—but prolonged or intense stresses can erode resilience [6].

Countermeasures are planned interventions designed to prevent or mitigate the negative impact of stressors. In spaceflight, countermeasures include exercise regimens to prevent bone loss, circadian lighting to stabilize sleep, private conferences with psychologists, and training in conflict resolution or self-regulation techniques [7].

Behavioral health encompasses the emotional, cognitive, and social well-being of astronauts. NASA defines it as the capacity to cope with mission stress, work productively, and maintain positive interpersonal relationships in the space environment. Behavioral health is monitored through pre-flight screening, in-flight psychological support, and post-mission reintegration programs [8].

Team performance describes how effectively a crew collaborates to achieve mission objectives. It is influenced by crew cohesion, communication quality, workload distribution, and conflict resolution skills. In space missions, where crews may be isolated for months, sustaining team performance is essential for safety and mission success [9].

3. Literature review

3.1 Cognitive and psychological stressors

Spaceflight exposes astronauts to a composite of environmental and operational stressors—defined earlier as environmental, operational, and psychological challenges to crew performance. These include prolonged weightlessness, confinement, disrupted circadian rhythms, elevated carbon dioxide levels, high noise, and radiation [2]. Each factor can impair cognition or mood: for example, elevated CO_2 and noise can degrade attention and decision-making [10]. Long missions also risk sleep deficiency, fatigue, and sensory monotony, which can slow reaction time and raise irritability. These conditions collectively create a "high-demand cognitive environment," where mental resources must be carefully managed—a challenge shared by other high-risk sectors such as nuclear power plant operations and air-traffic control.

Recent studies of ISS crews confirm that overall cognitive performance is largely maintained during six-month missions [11]. In one NASA study of 25 ISS astronauts, group averages showed no systematic decline in cognitive test scores compared to pre-flight baselines. However, transient changes occurred in specific domains: Early in flight, astronauts exhibited modest slowing in processing speed, visual working memory, and sustained attention, and late in flight, they showed a reduction in risk-taking propensity [11]. Notably, about 12% of in-flight or post-flight test scores fell 1.5 standard deviations or more below baseline, indicating that a subset of assessments showed the significant decrements [11]. These findings suggest that high-performing crews can generally maintain performance, but individuals may experience mission stage-dependent deficits.

Other research reports more pronounced cognitive impacts under the spaceflight conditions. For instance, Takács et al. [12] found that visuospatial performance deteriorated significantly during spaceflight. In their ISS study, astronauts were slower and more error-prone on spatial orientation tasks in orbit than on Earth, at both early (~8 days) and later (~50 days) mission stages. Brain imaging showed the diminished attentional resources, implying that even fundamental perceptual skills may suffer under microgravity and associated stress [12]. This finding has implications for the design of spacecraft displays and controls—ensuring that information is presented in a way that minimizes spatial disorientation.

Lessons learned here are increasingly being applied to other safety-critical systems on Earth, such as air-traffic control rooms and remote operations centers. Space missions also pose psychological hazards. Astronauts face enduring separation from family, intense workloads, and high responsibility. Surveys and reviews indicate the elevated rates of mood disturbances and anxiety among crew members. One analysis reported that a substantial fraction of astronauts experience subclinical or clinical symptoms during or after missions: NASA surveys found that 23–85% of astronauts (the range reflecting gender differences) reported anxiety symptoms, and 35–43% reported the depressive symptoms [10]. Moreover, almost all returning astronauts describe some cognitive or performance-related complaints linked to central nervous system effects [10]. The NASA "Twins Study" famously showed that a 12-month mission can induce neurocognitive and physiological changes that persist for months post-flight [10].

The composite nature of space stressors amplifies psychological risk. Microgravity, isolation and confinement, disrupted sleep, and mission stressors can cumulatively lead to fatigue, irritability, mood swings, and "space fog" (subjective cognitive slowing) [2, 10]. The Canadian Space Agency notes that isolation in space frequently leads to boredom, loneliness, and anxiety, and such feelings are expected to worsen in deep-space missions with the less contact with Earth [13]. In short, while many astronauts are highly resilient, the space environment can push coping limits. This insight has parallels on Earth: Shift workers in remote oil platforms, submariners, and Antarctic overwintering crews experience similar psychological stress profiles—a reminder that countermeasures developed for space can improve resilience in other extreme environments. Monitoring and supporting mental health is thus a mission imperative [13, 14].

3.2 Team dynamics and behavioral factors

Beyond individual stress, team performance (how effectively the crew collaborates to accomplish mission objectives) is pivotal. Small teams must cooperate closely for months or years in isolated settings. Differences in background, language, and culture can lead to misunderstandings and tension. Historical accounts highlight these psychosocial challenges: For example, astronaut Norman Thagard's record-setting 115-day Mir mission in 1995 was marked by communication difficulties, cultural differences with his Russian hosts, heavy workload, and sheer duration—he reported these as major stressors [15]. Space agency analyses similarly note that factors such as lack of privacy, cramped quarters, and interpersonal conflicts can degrade morale and performance [13, 15].

Team performance may also degrade over time, particularly in long-duration missions. Analogs and expedition studies (e.g., Mars-500, Antarctic winter-overs) often report increases in irritability and group friction as stress accumulates. Experts emphasize that training in conflict management and emotion regulation is essential before flight [16]. For instance, European Space Agency analysts recommend that crews be trained to recognize and cope with conflicts or mood shifts on their own, using structured debriefing and coping strategies [16]. NASA's behavioral health program (focused on astronauts' psychological well-being) likewise ensures that astronauts receive extensive team-skills training and practice effective group-living techniques to build teamwork and conflict-resolution abilities [14].

Social isolation from Earth is another significant challenge. Even with the modern communication, time delays in deep-space missions will further isolate astronauts.

Maintaining meaningful ties to home—through delayed video calls, emails, and virtual "care packages"—helps support morale [13, 14]. Current ISS expeditions include countermeasures (interventions to counteract stress and risks) such as regular private conferences with psychologists and crew journaling, which have proven helpful in managing stress [14]. However, beyond low Earth orbit these supports will be limited by distance, heightening the importance of robust interpersonal cohesion and self-monitoring within the crew. These findings echo best practices in high-stakes terrestrial environments, where fostering psychological safety and communication protocols among small teams is critical to preventing errors.

3.3 Decision-making under autonomy and uncertainty

As missions venture farther, crew autonomy in decision-making will increase out of necessity. Earth-based flight controllers have traditionally guided ISS crews for many tasks, but communications delays—which may reach several minutes for Mars missions—will require crews to make critical decisions without real-time input from mission control. Marquez et al. [17] studied "self-scheduling" in high-fidelity analog missions, showing that giving crews the ability to adjust their own schedules is feasible but also raises workload. They concluded that as crews move to future exploration missions, they will need to operate more autonomously due to the delayed and intermittent communications [17]. Without continual Earth oversight, astronauts must become adept at dynamic planning and *in situ* problem solving.

This shift brings risks: Removing ground support can quickly overwhelm crews if not properly managed. Marquez and colleagues found that if operational assistance is reduced without compensating training and decision-support tools, crews may face higher workload and increased errors [17]. The NASA human factors community also note that greater crew autonomy has potential downsides: If not designed carefully, astronauts may either over-rely on flawed automation or make inconsistent decisions without external checks. For example, on the ISS, automated control scripts were introduced to reduce human command errors, but it was observed that automation can induce complacency and reduce human situational awareness [4]. This "automation complacency" highlights the need to design autonomy systems that keep crews appropriately engaged and informed.

The lessons learned from autonomy research in space are also valuable for terrestrial industries. Remote operations in nuclear power, maritime shipping, and deep-sea oil drilling face similar challenges: Balancing automation with human oversight to prevent over-reliance and ensure operators maintains situational awareness. Designing decision-support tools that are transparent, intuitive, and adaptable is a shared priority across these sectors.

Overall, decision-making in space is challenged by uncertainty and stress. Cognitive limitations under fatigue and pressure can impair judgment. Without immediate expert support, even routine decisions (such as scheduling experiments or responding to minor anomalies) become weighty. Human factors research emphasizes that error-tolerant procedures, checklists, and realistic simulations are the vital supports for autonomous crews. In deep-space missions, "resilience by design" means anticipating that errors will occur and embedding recovery paths into every operational procedure. Harris and Simpson [4] recommend a twofold approach for ISS operations: implement human-centered design to minimize opportunities for error, and simultaneously assume errors will occur and design robust recovery procedures into every process [4]. Such resilience-by-design thinking—accepting human fallibility and planning around it—will be essential on autonomous deep-space missions.

4. Human: System integration and error risk

"Humans will make errors" is a core axiom of human-systems engineering. The complexity of space operations, combined with spaceflight stressors, guarantees that some errors will occur. On the ISS, one NASA analysis found that 57% of serious mission incidents (1996–2005) involved human error, echoing trends in other high-risk domains where the majority of incidents have human causes [4]. In practice, mitigating human error on the ISS has involved iterative refinement of tools and procedures. For instance, NASA's Payload Operations team uses pre-programmed command scripts and automated check systems to catch common mistakes. They also employ standardized procedures, streamlined information flow, and rapid crew feedback loops—coupled with extensive training—to reduce errors [4].

Space agencies today advocate a philosophy of human-systems integration (HSI): Incorporating human capabilities and limitations at every stage of system design. NASA's updated risk management guidance explicitly links system resilience to human factors, arguing that designing crew health and performance systems through rigorous HSI processes improves overall mission outcomes [1]. In other words, spacecraft and habitats should provide a "human margin"—spare cognitive and physical capacity to handle off-nominal events [2]. This includes not only physical habitat layout (adequate living space, privacy) but also interface design (clear displays, augmented-reality aids, automation with transparent status and warnings), as well as robust training and support. Recognizing that eliminating all human error is unrealistic [4], mission designers instead strive to make systems tolerant of human fallibility: Tasks are structured with cross-checks, critical operations are double-verified, and crews are trained in error detection and recovery.

Human-systems integration in space must go beyond ergonomics and account for microgravity-specific constraints, such as tool tethering, interface orientation, and the cognitive effects of CO_2 buildup or circadian rhythm disruption. These requirements are not directly transferable from terrestrial HSI frameworks and must be validated through space-specific testing.

The methodologies refined in space HSI programs have influenced terrestrial safety-critical sectors—for example, aviation cockpit design and nuclear control room interfaces. Conversely, advances in human factors from these Earth-based industries feed back into space design, creating a two-way exchange of resilience knowledge.

5. Resilience strategies and countermeasures

Given these human performance vulnerabilities, mission planners are shifting from expecting fault-free operation to deliberately building resilience. In this context, resilience refers to the ability of a crew and their systems to anticipate, withstand, and recover from stressors or operational errors without catastrophic mission failure. Space agencies are exploring multiple strategies to enhance resilience at both individual and system levels. Key focus areas include crew selection and training, in-flight support, environmental and procedural design, and the use of Earth analogues to model and mitigate risks (**Table 1**).

Countermeasure category	Examples and approaches
Crew selection and training	Rigorous screening of astronaut candidates for adaptability, stress tolerance, and teamwork skills. Extensive pre-flight training in technical competencies as well as psychological readiness (e.g., stress inoculation, conflict resolution, cultural competency). Astronauts are taught self-care techniques and provided tools (relaxation strategies, sleep management aids) to cope with stress and maintain mental health during missions [14]. Emerging approaches include adaptive e-learning platforms and immersive virtual reality team training, which can be updated mid-mission to address unforeseen challenges.
In-flight support	Ongoing measures to sustain mental well-being during missions. Current ISS practices include regular private video conferences with psychologists, confidential mood monitoring, structured exercise and recreational activities, and opportunities for communication with family [13, 14]. Crews also use personal journals and scheduled debriefs to process stress and interpersonal issues. For future deep-space missions with limited communication, researchers are developing more autonomous support tools (onboard health monitoring systems and AI-based "virtual counselor" agents) to assist crews without immediate Earth contact. This mirrors the developments in telemedicine and AI-driven mental health support on Earth, such as digital CBT platforms used in remote military deployments.
Environmental and procedural design	Design of spacecraft and habitats to proactively manage stressors. Long-duration transit vehicles and surface habitats provide private sleeping quarters, exercise facilities, and social spaces to mitigate monotony and isolation. Circadian lighting systems and noise control help stabilize sleep cycles. Work schedules incorporate adequate rest and allow some crew self-scheduling flexibility to prevent fatigue [17]. Critical tasks are planned with margins to avoid cognitive overload. NASA and ESA are increasingly incorporating human-centered design reviews into every phase of habitat development, similar to human-systems integration processes in aviation and nuclear power industries.
Ground analogs and simulations	Extensive use of Earth-based analogue missions (e.g., Antarctic stations, NEEMO undersea habitat, HI-SEAS, Mars-500) to study crew behavior and test countermeasures under the conditions of isolation and confinement. Analog crews practice autonomy and conflict resolution, often with structured debriefs to improve team functioning [16]. These simulations also enable testing of new technologies (such as virtual reality "nature" experiences for stress relief) and training scenarios involving delayed communication, helping to prepare crews for high-autonomy deep-space missions. The data from these analog missions is now being used not only for spaceflight preparation but also for optimizing operations in remote terrestrial research stations and Arctic energy installations.

Table 1.
Resilience countermeasures for human spaceflight.

6. Discussion

Bringing together these findings, it is clear that the human element is both a critical asset and a potential Achilles' heel of space missions. On one hand, well-trained, adaptable humans can solve unexpected problems and make flexible judgments beyond current AI capabilities. On the other hand, human physical and psychological limits impose constraints on mission design and operations. Resilience must therefore be engineered at both the human and system levels to ensure mission success.

Despite advancements in countermeasures, significant knowledge gaps remain. Much of the existing research focuses on missions of relatively short duration (6 months or less) with continuous Earth support, whereas Mars expeditions will be far longer and more isolated [2]. Few studies have systematically analyzed cognitive and psychological trajectories over missions beyond 1 year. Many countermeasures (e.g., mindfulness training, group therapy techniques, exercise regimens) have been tested only in analog environments, with mixed evidence of their long-term efficacy in actual spaceflight [2]. There is also limited knowledge on optimal crew composition: NASA is investigating which combinations of personalities and skills yield the most cohesive teams, but this remains an open question. In addition, the interactions of multiple deep-space stressors (radiation, isolation, microgravity, etc.) on the brain and behavior are only beginning to be understood [10].

Several practical limitations in the current mitigation approaches can be identified. First, many strategies assume Earth-like support that will not exist beyond low Earth orbit. For example, ISS crews can receive real-time psychological counseling or even evacuate in an emergency; future crews on Mars must be self-reliant for extended periods. Yet today's astronaut training is still heavily front-loaded on preflight learning. The NASA Human Research Program has highlighted the need for "adaptive in-mission training" to refresh skills during flight and prevent performance decay [7]. This could include augmented-reality guided procedures, peer-to-peer scenario practice, and even just-in-time skill refreshers—an approach that has parallels with continuing professional education requirements in medicine, aviation, and energy control-room operations.

Second, most countermeasure research and protocols have not fully accounted for cultural and individual differences. Many space analog studies involve relatively homogeneous crews; however, future Mars teams may be more diverse in nationality, culture, and gender. Cultural training and language preparation will become more important, as will strategies for conflict resolution in multicultural groups [16]. Gaps remain in understanding how stress and behavioral health issues manifest across different personality types and between genders; for instance, some evidence suggests female astronauts report anxiety symptoms at higher rates than males [10]. Psychological support protocols should be inclusive and validated for all demographics.

Third, technological solutions (automation, robotics, telemedicine) hold great promise but also carry their own risks. Over-reliance on automation, for example, can erode situational awareness [4]. Designing effective human-robot teams is non-trivial, and best practices are still emerging [1]. Interfaces and decision-support tools must be developed so that astronauts can intuitively manage complex systems without confusion or over-trust. In essence, advanced technology should assist—not replace—human decision-making, and it must be introduced with careful attention to human factors to avoid new failure modes. This concern is echoed in civilian domains, where over-automation in aviation has been linked to loss of manual flying skills and slower anomaly detection.

7. Recommendations

Based on the above analysis, several recommendations can help bolster human resilience in future space exploration missions:

- *Develop adaptive in-flight training regimens.* Space agencies should implement continuous learning opportunities for crews during long missions. For example, interactive digital tutors or augmented reality systems could provide just-in-time refreshers of critical procedures. Training should emphasize transferable cognitive and teamwork skills (problem-solving, leadership, situational awareness) rather than rely solely on rote memorization of procedures [7]. In ground-based analogs, crews should practice self-directing schedules and tasks to build autonomy and decision-making confidence [17]. Similar training refresh systems have been adopted in high-reliability industries such as nuclear energy and commercial aviation, providing a cross-sector validation of this approach.

- *Enhance team composition strategies and behavioral health screening.* Astronaut selection should place greater weight on traits linked to resilience—such as emotional stability, adaptability, and strong interpersonal skills. Ongoing research into "team chemistry" should be expanded by collecting data from both analog missions and space crews to inform how diverse teams function over time [2]. Pre-flight team building and cultural competence training should be strengthened, including workshops on conflict management and communication [16]. These lessons are transferable to Earth-based mission control and crisis-management teams, where psychological compatibility and trust are also performance-critical. *Prioritize human-centered habitat and spacecraft design.* Spacecraft and habitat architectures must treat human needs as central design criteria. This entails providing sufficient living space, privacy, and leisure opportunities to mitigate stress and isolation [2]. Environmental control systems (lighting, temperature, noise) should support circadian rhythms and comfort. Critically, automation and decision-support tools should be developed through iterative user testing with astronauts to achieve the right balance of trust. Crews should be able to rely on autonomous systems, yet maintain situational awareness and the ability to override when necessary [4]. All interfaces must be user-centered, with clear feedback to prevent mode confusion and automation complacency.

- *Invest in autonomous behavioral health support tools.* In preparation for reduced Earth contact, space agencies should accelerate the development of onboard mental health resources. For example, virtual reality applications could allow astronauts to experience calming "Earth-like" environments for stress relief. Conversational AI "counselor" programs might deliver cognitive-behavioral therapy exercises or simply provide an empathetic ear during long-duration missions. Biosensors and software could continuously monitor indicators of stress or depression (such as sleep quality and heart rate variability) and confidentially alert the crew or support teams to emerging issues. Such approaches are already being tested in remote healthcare settings on Earth, where mental health resources are scarce. *Integrate human factors into mission planning and policy.* Mission timelines and objectives should include margins for crew rest and adaptation. Planners must avoid overloading crews with excessive tasks at the expense of recovery time. Space agencies should mandate formal human factors and behavioral health reviews as part of mission design, alongside engineering safety reviews. International collaboration in research—sharing data and best practices among NASA, ESA, and other agencies—will be vital for addressing these challenges. Finally, funding should support more long-duration isolation studies (including multi-year analog missions) to better simulate and understand the psychological demands of interplanetary travel. Similarly, operators of terrestrial critical infrastructure—such as national grids and air-traffic networks—could benefit from adopting human-centered workload margins as standard safety practice.

8. Broader implications: Resilience lessons for earth

The resilience strategies explored in this chapter do not end at the edge of Earth's atmosphere. Space missions depend on a network of terrestrial facilities—mission control centers, communications hubs, and ground stations—where human operators work in high-pressure, around-the-clock environments. Just as in space, human error or fatigue in these settings can have cascading consequences for safety and mission success.

Designing Earth-side operations with a "resilience by design" approach is therefore critical. Examples include staffing schedules at mission control that allow sufficient rest, redundant communication links and procedures to prevent single points of failure, and human-centered software interfaces that help ground operators maintain situational awareness during anomalies. These practices are now being studied jointly by NASA, ESA, and international partners, reflecting a recognition that space operations and terrestrial critical infrastructure share common vulnerabilities.

Moreover, insights from astronaut behavioral health programs are informing resilience planning for sectors such as aviation, nuclear power, and cybersecurity operations. Structured team training, error-tolerant procedures, and real-time psychological monitoring are gradually becoming part of the safety culture in these industries. The flow of knowledge is bi-directional: lessons learned from air-traffic controllers' workload management and from nuclear control-room HSI reviews are feeding into the design of next-generation lunar mission operations centers.

By reframing this section as a forward-looking perspective, we underscore that the lessons of human spaceflight have the potential to strengthen resilience far beyond the space domain—an important point for policy-makers, system designers, and operators of critical infrastructure on Earth.

9. Conclusion

The human element is the keystone of mission success and safety in twenty-first-century space exploration. Astronauts must function not only as engineers and scientists, but also as stewards of their own mental health and team dynamics. This chapter reviewed the multifaceted risks—cognitive strains, psychological stress, group dynamics, and operational errors—that can threaten long-duration missions. Current research from ISS crews and analog simulations provides valuable insights but also reveals critical gaps. Cognitive performance may remain mostly stable for months, yet subtle impairments and a high mental health burden loom on the horizon of multi-year missions [10, 11]. Increasing crew autonomy and reliance on technology introduce new modes of potential failure that require fresh design solutions [4, 17].

Importantly, human-centered design and resilience-building strategies can mitigate many of these risks. Space agencies already employ robust selection, training, and in-flight support programs, but these must be extended and innovated for the unprecedented challenges of Mars-class missions. By integrating behavioral health considerations into every mission phase—from spacecraft design to daily schedules—we can create conditions where astronauts are not only kept safe but can thrive even under extreme conditions. This will involve continued innovation: adaptive in-flight training, autonomous mental health tools, and habitat designs that holistically support both physical and psychological well-being.

Ultimately, space exploration will always carry inherent risks. Yet by keeping humans at the center of system design, and by cultivating psychological resilience as deliberately as we engineer life-support systems, we can significantly enhance mission safety and performance. The lessons of space human factors now form a valuable toolkit for designing safer, more resilient critical infrastructures on Earth—from power grids and transport networks to cybersecurity operations. Future crews, whether in orbit or in mission control rooms on Earth, should be "resilient by design."

Author details

Ulpia-Elena Botezatu[1*] and Victor Vevera[2]

1 National Institute for Research and Development in Informatics, ICI Bucharest, and Romanian Space Agency (ROSA), Romania

2 National Institute for Research and Development in Informatics, ICI Bucharest, Romania

*Address all correspondence to: ulpia.botezatu@ici.ro

References

[290] Antonsen EL, Connell E, Anton W, Reynolds RJ, Buckland DM, Van Baalen M. Updates to the NASA human system risk management process for space exploration. npj Microgravity. 2023;**9**(1):72. DOI: 10.1038/s41526-023-00305-z

[291] Pagnini F, Manzey D, Rosnet E, Ferravante D, White O, Smith N. Human behavior and performance in deep space exploration: Next challenges and research gaps. npj Microgravity. 2023;**9**(1):27. DOI: 10.1038/s41526-023-00270-7

[292] Patel ZS, Brunstetter TJ, Tarver WJ, Whitmire AM, Zwart SR, Smith SM, et al. Red risks for a journey to the red planet: The highest priority human health risks for a mission to Mars. npj Microgravity. 2020;**6**(1):33. DOI: 10.1038/s41526-020-00124-6

[293] Harris SS, Simpson B. Human error and the International Space Station: Challenges and triumphs in science operations. In: Proceedings of the AIAA Aviation Technology, Integration, and Operations Conference (AIAA 2016-3763). 2016. Available from: https://ntrs.nasa.gov/api/citations/20160006994/downloads/20160006994.pdf

[294] National Academies of Sciences, Engineering, and Medicine. A Midterm Assessment of Implementation of the Decadal Survey on Life and Physical Sciences Research at NASA. Washington, DC: The National Academies Press; 2018. DOI: 10.17226/24966

[295] Stuster J. Bold Endeavors: Lessons from Polar and Space Exploration. Annapolis, MD: Naval Institute Press; 2011

[296] NASA Human Research Program. Human Research Roadmap: Behavioral Health and Performance Risk Mitigations and Gaps. 2023. Available from: https://www.nasa.gov/hrp/

[297] Kanas N, Manzey D. Space Psychology and Psychiatry. 2nd ed. Dordrecht: Springer; 2008. DOI: 10.1007/978-1-4020-6770-9

[298] Salas E, Fiore SM, Letsky M. Theories of Team Cognition: Cross-Disciplinary Perspectives. New York: Routledge; 2011

[299] Yin Y, Liu J, Fan Q, Zhao S, Wu X, Wang J, et al. Long-term spaceflight composite stress induces depression and cognitive impairment in astronauts-insights from neuroplasticity. Translational Psychiatry. 2023;**13**(1):342. DOI: 10.1038/s41398-023-02638-5

[300] Dev SI, Khader AM, Begerowski SR, Anderson SR, Clément G, Bell ST. Cognitive performance in ISS astronauts on 6-month low earth orbit missions. Frontiers in Physiology. 2024;**15**:1451269. DOI: 10.3389/fphys.2024.1451269

[301] Takács E, Barkaszi I, Czigler I, Pató LG, Altbäcker A, McIntyre J, et al Persistent deterioration of visuospatial performance in spaceflight. Scientific Reports. 2021;**11**(1):9590. DOI: 10.1038/s41598-021-88938-6

[302] Canadian Space Agency. How Space and Isolation Affect Astronauts' Mental Health. Government of Canada; 2022. Available from: https://www.asc-csa.gc.ca/eng/youth-educators/toolkits/mental-health-and-isolation/how-space-and-isolation-affect-astronauts-mental-health.asp

[303] Lewis RE. Behavioral health. In: NASA Human Health and

Performance Directorate. 2023. Available from: https://www.nasa.gov/hrp/human-health-countermeasures/

[304] Herring L. Astronaut draws attention to psychology, communication. Human Performance in Extreme Environments. 1997;**2**(1):42-47

[305] De la Torre GG, Groemer G, Diaz-Artiles A, et al. Space Analogs and Behavioral health performance research review and recommendations checklist from ESA topical team. npj Microgravity. 2024;**10**:98. DOI: 10.1038/s41526-024-00437-w

[306] Marquez JJ, Hillenius SM, Kanefsky B, Zheng J, Deliz I, Reagan M, et al. Increasing crew autonomy for long duration exploration missions: Self-scheduling. In: Proceedings of the Human Systems Integration Conference. NASA Ames Research Center; 2017

Chapter 9

Perspective Chapter: Pioneering Analog Astronaut Research in Africa – The OASEAS Habitat in Kenya

Isaac Macharia

Abstract

Africa is rising as a hub for space exploration and research, with the continent increasingly embracing the challenge of fostering innovation and addressing global scientific frontiers. Within this context, Kenya stands at the forefront, leveraging its unique geographical, cultural, and intellectual assets. Spearheading these efforts is Mars Society Kenya (MSK), a visionary organization advancing the boundaries of analog astronaut research. Central to MSK's initiatives is the Omni-Africa Space Exploration Analog Simulation (OASEAS) habitat, an ambitious project set to establish a cutting-edge research station in the Chalbi Desert of Marsabit County, Kenya. This project not only exemplifies the study of Mars' physical environment but also Kenya's commitment to space science, while positioning Africa as a critical player in global space exploration.

Keywords: Kenya, Africa, space, analog, habitats

1. Introduction

The African space economy is undergoing a transformative shift marked by institutional advancements and technological growth. This shift is reshaping how African nations engage with space, aiming to leverage it for socio-economic development [1].

Kenya is emerging as a significant player in space science and research, which is marked by significant milestones that highlight its growing role in the field. The establishment of the Kenya Space Agency (KSA) in 2017 has been a cornerstone [2, 3], enabling coordinated efforts in satellite launches, research, and international collaboration.

At the forefront of this transformation is Mars Society Kenya (MSK) [4, 5]; a not-for-profit organization legally registered in Kenya as a Non-Governmental Organization (NGO) under the Kenya's Public Benefit Organizations (PBO) Act of 2013, which was fully operationalized in May 2024 under Legal Notice No. 78. A key part of the Mars Society Kenya initiative is the development and operation of an Analog Habitat Research Station in the Chalbi Desert of Marsabit County. Named the Omni-Africa Space Exploration Analog Simulation (OASEAS), this habitat serves as a critical platform for its members to engage in practical, hands-on experiences that contribute to the global understanding of extraterrestrial habitation (Mars Society Kenya – IP documentation).

This initiative not only propels Africa into the global space research arena but also pioneers analog astronaut research on the continent. By simulating extraterrestrial conditions in the challenging environment of the Chalbi Desert, OASEAS will enable researchers in Africa to participate in space science, engineering and technologies such as: robotics, energy, water and waste management, architecture, biotechnology, agri-technology, food technology. As the first private and sub-Saharan African habitat in a global ecosystem of analog space research stations such as the HiSEAS in Hawaii, LunARES in Poland, and the Mars Desert Research Station (MDRS) in Utah [6], the OASEAS will also address the challenge of advancing space for all humanity by advancing transnational research partnerships such as those envisaged in the Artemis Accords.

Over 370 million Indigenous individuals inhabit 90 countries across all continents. They are the speakers of approximately two-thirds of the world's 7000 living languages. While there is no single, universally agreed-upon definition of *"Indigenous peoples,"* the United Nations Permanent Forum on Indigenous Issues (UNPFII) describes them as those who identify themselves as Indigenous, are recognized as part of Indigenous communities, and maintain ties to the societies, lands, languages, traditions, and governance systems that existed before colonial or settler influences. These communities are committed to preserving their ancestral lands and distinct ways of life for future generations ([7], para.)

Indigenous communities in Africa offer a rich tapestry of knowledge, resilience, and adaptive strategies that can serve as meaningful parallels and lessons for space analog research. Indigenous groups are characterized by their enduring connections to pre-colonial societies on their territories. They see themselves as separate from the dominant cultures that now occupy those regions. Often forming marginalized segments of modern society, they strive to safeguard and pass down their lands, identities, and cultural practices, rooted in their unique social, legal, and cultural frameworks [8]. Space analog environments aim to mitigate the challenges of extraterrestrial isolation, including resource scarcity and space debris [9]. Many African indigenous groups have thrived for centuries under similarly harsh circumstances, developing sustainable practices and social systems that could inform how humans prepare for and live in space.

2. Africa's ascent into SMART futures

The OASEAS habitat in Kenya represents a groundbreaking step in Africa's journey toward becoming a key player in space exploration. By integrating scientific research with cultural knowledge, addressing local and global sustainability challenges, and inspiring the next generation of African space scientists, OASEAS is not just for running analog mission—it is a blueprint for Africa's contribution to humanity's future [10, 11]. As the project continues to evolve, it will serve as a model for how space research can be both globally impactful and locally transformative, ensuring that Africa's voice is heard in the quest to explore and inhabit new worlds.

Kenya has been a hub for space activities since hosting the Luigi Broglio Space Centre, facilitating satellite launches and research. Strategically positioned near the Equator, Kenya offers an ideal launch site—cheaper and more efficient due to the Earth's rotational boost. This geographic advantage could transform the continent into a global hub for spacefaring nations and private companies alike, driving economic growth through tech innovation, job creation, and infrastructure development [12].

But the ripple effects go deeper. Kenya's focus on space-based solutions—monitoring climate change, improving agriculture, and enhancing connectivity—addresses Africa's immediate challenges while building expertise for the future. In future, a network of African satellites could provide real-time data to combat desertification or predict floods, turning a continent often seen as a victim of global crises into a leader in resilience. Over decades, this could position Africa as a key player in sustainable space technologies, exporting knowledge to the world and redefining its economic narrative from resource extraction to intellectual capital.

Africa's emerging space industry connects these threads. Countries like Nigeria, South Africa, and Kenya are advancing satellite technology [13], while the African Union promotes local innovation. Indigenous knowledge—like Ethiopia's ancient stargazing heritage—could propel these efforts, merging tradition with modernity as humanity looks to the Moon and Mars. Recognizing Indigenous identity goes beyond definitions—it is about respecting the names they choose for themselves and their lands, rather than those imposed by modern states.

International bodies, like the United Nations through its Declaration on the Rights of Indigenous Peoples (UNDRIP), affirm that honoring Indigenous knowledge and practices supports sustainable development and environmental care. In recent years, Indigenous communities have pushed for rights to self-governance, land, culture, and education, with global resolutions reinforcing their role in ecological stewardship [14].

2.1 Cultural context and local integration

As Mars Society Kenya continues to advance its mission at the Omni-Africa Space Exploration Analog Simulation (OASEAS), they find themselves at a pivotal moment. Our journey into the cosmos is guided by international law [15]. It is not just a testament to our technological prowess but also a reflection of our collective spirit and values by integrating the spirit of *Ubuntu* into the management of OASEAS [16–18]. *Ubuntu*, an African philosophy that emphasizes community, shared humanity, and mutual care, can profoundly enhance the project's success and significance.

Ubuntu is encapsulated in the phrase, "I am because we are," highlighting the interconnectedness of all people. It champions the idea that our individual well-being is inextricably linked to the well-being of others. This philosophy promotes empathy, cooperation, and a sense of shared purpose, which are crucial for any collaborative endeavor. Culturally, African Indigenous traditions often emphasize community and collective decision-making. The *Ubuntu* philosophy, which values interconnectedness, could enrich space missions. In the tight quarters of a spacecraft, where teamwork is vital, *Ubuntu*-like principles might improve crew harmony, conflict management, or mission design. Space agencies already explore group dynamics for long missions—adding such cultural insights could strengthen cohesion [19].

Specific traditions also offer value. West African griots preserve history through oral storytelling, a practice that could boost morale or share knowledge in space, supporting mental well-being. Similarly, the use of local plants like Artemisia for healing in East Africa reflects a resourcefulness that might encourage astronauts to experiment with extraterrestrial materials for medical needs, reducing dependence on Earth.

Across various African regions, traditional authorities enforce resource conservation through strict penalties. Offenders who break these rules might face fines, which can vary widely: in Ethiopia, penalties might include cash payments or, in severe cases, exclusion from community services. In Zimbabwe, violators may owe livestock or grain to local leaders, while in Ghana, punishments range from monetary fines or livestock offerings to public flogging or even banishment from the village. The fear of these consequences has proven effective in safeguarding both plant and animal life.

In Zambia, the Tonga people have long upheld a custom, still observed today, where traditional leaders prohibit cutting down fruit-bearing trees for firewood. This practice reflects a deep respect for natural resources and their sustainable use. Likewise, in the Mount Cameroon area, village elders enforce rules that protect biodiversity and support community well-being. For example, outsiders are barred from collecting medicinal plants or logging timber from local forests. These leaders also promote planting trees with both healing properties and economic benefits, aiding in the preservation of key animal species like elephants, chimpanzees, bushbucks, and galagos, as well as plant species such as raffia palms and cordia trees [20].

One of the most compelling aspects of OASEAS is its cultural integration. Unlike other analog research projects, which often take a purely technological approach, OASEAS considers the cultural and social dynamics of the local communities in Kenya. The project is informed by indigenous knowledge on resource management, survival strategies in arid landscapes, and traditional ecological practices, offering insights into sustainable living both on Earth and in space.

Engaging with communities around Marsabit County fosters knowledge exchange between scientists and local experts, bridging modern technological advancements with centuries-old survival techniques. This approach not only enriches space research but also ensures that the project has immediate benefits for local populations, such as improving water conservation methods and developing efficient waste recycling systems.

It is so unfortunate that the United States and Europe dominate conversations around the space economy and this is because both countries have colonized the entire world. However, it is worth noting that way before Africans were colonized, we had our own ways of telling seasons through observing the sky and had our own stories to interpret what we saw. A good example is the stone circle of Nabta Playa which is the world's first and the oldest known astronomical site on Earth. It was built in Africa and is older than Stonehenge. This 7000-year-old stone circle tracked the summer solstice and the arrival of monsoon rains in the Sahara Desert thousands of years ago.

Many sites evidence the existence of African astronomy and cosmology, dating back to civilizations that meticulously tracked celestial bodies and integrated them into their lives. This journey is deeply connected and wonderful, with megalithic structures and cultural beliefs. Africa's astronomical legacy extends to its societies, calendars, and mythology, and continues to be shaped by African astronomers, who continue to push the boundaries of our understanding of the universe through cutting-edge research and international collaboration.

Traditional African storytelling is a cornerstone of the continent's cultural heritage, with its roots deeply embedded in communal life and its practices uniquely adapted to different regions and contexts. It transcends mere entertainment, acting as an educational tool, a vehicle for preserving history, and a framework for societal cohesion. Traditional African storytelling, when viewed through the lens of astronomy and space exploration, reveals a profound connection between cultural narratives and humanity's understanding of the cosmos. Many African societies used storytelling to explore celestial phenomena, explain the origins of the universe, and convey a deep respect for the stars and planets as integral to human life. These narratives often blended mythology, spirituality, and observational science, showcasing a sophisticated appreciation of astronomy long before modern scientific frameworks emerged.

Traditional African storytelling bridges the past and future, demonstrating humanity's timeless fascination with the universe. By connecting the oral traditions of African societies to contemporary astronomy and space exploration, these stories not only preserve cultural identity but also inspire new generations to reach for the stars.

2.2 Africa's growing influence in global space exploration

Africa's influence in global space exploration is accelerating [21–26]. Beyond Kenya, nations like South Africa (with its Square Kilometer Array) and Nigeria (a pioneer in satellite launches) are amplifying the continent's voice. Kenya's Malindi Space Centre, hosting an Italian-managed launch facility [27], already signals Africa's strategic importance. As the global space economy balloons—projected to hit $1 trillion by 2040—Africa's low-cost launch capabilities and growing talent pool could attract players like SpaceX or Blue Origin [28], shifting the industry's center of gravity southward [29].

By prioritizing affordable, practical technologies—like CubeSats for education or disaster management—African nations might challenge the dominance of wealthy superpowers and corporations. This is not just influence; it is disruption, positioning Africa as a moral and technical counterweight in a field long dominated by a few.

In Africa, Indigenous knowledge frequently centers on sustainability and adaptability, skills refined in diverse and challenging environments. For example, the San people of southern Africa have mastered survival in dry regions by interpreting faint natural clues to locate water. Such resourcefulness could influence space exploration, where essentials like water, air, and food are scarce. Their methods might inspire efficient life-support systems or recycling techniques in space. For example, using their approach to pinpoint scarce resources in alien soils or ice—a blend of practical, low-tech wisdom with advanced technology.

For millennia, Indigenous communities worldwide have tailored their practices to their environments, fostering sustainable livelihoods. In the Andes, for instance, agricultural techniques dating back 8000 to 10,000 years—like rotational farming, shifting cultivation, pastoralism, fishing, agroforestry, and hunting—demonstrate this enduring commitment to ecological balance [30].

Navigation is another strength. Certain African groups, such as the star-savvy navigators of Madagascar or the Dogon people, who possess intricate astronomical knowledge, have long used the night sky to guide them. The Dogon's understanding of the Sirius star system, potentially even its faint companion, predates modern astronomical tools. This expertise could enhance space travel, offering astronauts alternative ways to navigate beyond Earth's GPS-reliant zones, bolstering safety alongside high-tech systems. Their frameworks for understanding the cosmos could inspire alternative perspectives on space exploration, blending practical survival with a sense of purpose—something vital for long-duration missions where morale can falter.

Analog research is a critical step in preparing humanity for interplanetary missions. While such research has been extensively conducted in regions like North America and Europe, Africa's involvement remains limited. The establishment of

OASEAS marks a significant milestone by providing an indigenous platform for testing space technologies in Mars-like conditions.

By leveraging Africa's unique environmental diversity, OASEAS is positioned to conduct high-impact research on human adaptation to extreme conditions, sustainability in resource-limited environments, and the practical application of space technologies for Earth-based challenges. Furthermore, this initiative underscores Africa's capability to contribute to global space exploration efforts by offering a distinct perspective informed by local knowledge systems and traditional resilience strategies.

2.3 Human life on other planets

Africa's space efforts could shape how humans live beyond Earth. Kenya's experience with arid climates and resource scarcity mirrors the challenges of extraterrestrial habitats like Mars. The continent's innovations—say, drought-resistant crops or solar-powered systems—could become blueprints for sustaining life on other planets. In future, we could have a Martian colony powered by Kenyan-engineered microgrids or fed by farming techniques honed in the Sahel.

Moreover, Africa's collaborative ethos could influence the social fabric of off-world settlements. Unlike the competitive space race of the twentieth century, Africa's approach—seen in partnerships like the African Union's space strategy—emphasizes unity. This could foster a model for planetary governance that prioritizes equity over-exploitation, a stark contrast to dystopian visions of space as a corporate free-for-all. Africa might not just send astronauts; it could export a philosophy of coexistence [31–33] critical for humanity's survival in the cosmos.

Despite challenges, these communities demonstrate remarkable adaptability. In Tanzania, the Hadza have engaged in carbon offset projects to protect their woodlands, while in Namibia, Indigenous conservancies have tripled elephant populations through community-led efforts. Such examples illustrate how remote living can foster agency when supported by policies that respect self-determination and traditional governance systems.

Traditional Ecological Knowledge (TEK) embodies the wisdom, customs, and beliefs about nature that indigenous and local groups have cultivated over generations through close ties to their environments. For many African indigenous communities, TEK is woven into their cultural identity, spiritual practices, and daily survival, showcasing an intimate understanding of local ecosystems, weather cycles, and resource stewardship.

Many African communities also maintain Sacred Natural Sites (SNS), which serve as spiritual centers for worship and rituals. These sites—ranging from sacred groves and church forests to hills, caves, and rivers—vary in size from small patches to vast expanses, often standing out in the landscape. These places are thought to house nature spirits or deities that oversee and protect the land and its resources. Community custodians regularly perform ceremonies, prayers, and offerings at these sites to honor or calm these spirits, ensuring their continued presence and the vitality of these sacred areas [34].

The Himba of northern Namibia, a semi-nomadic pastoralist group, often live in small, tight-knit communities far from urban centers. Their social cohesion, maintained through storytelling, rituals, and shared responsibilities, helps them endure physical and psychological isolation [35]. Space crews, confined for months or years, could draw lessons from these practices to bolster mental health and group dynamics. The Himba's conflict resolution methods, often mediated through communal dialog, might even offer a low-tech blueprint for managing interpersonal tensions in confined habitats.

The impact of OASEAS extends beyond its immediate research objectives. This project serves as a catalyst for STEM (Science, Technology, Engineering, and Mathematics) education in Africa, inspiring young scientists, engineers, and innovators to engage in space-related fields. By demonstrating that Africa can contribute meaningfully to space research, OASEAS fosters a new generation of space scientists and entrepreneurs who can drive the continent's space agenda forward.

Additionally, the project has broader implications for Africa's industrial and technological advancement. The solutions developed in the OASEAS habitat, particularly in waste recycling, water purification, and sustainable agriculture, have direct applications for addressing challenges such as desertification, water scarcity, and urban waste management. The technological advancements and innovations emerging from this initiative have the potential to benefit not only Africa but also humanity's broader quest for sustainable living, both on and off Earth.

2.4 Africa's role in humanity's future beyond earth

As humanity looks toward colonizing other planets, Africa's contributions through projects like OASEAS become increasingly relevant. The lessons learned from operating in extreme environments, coupled with indigenous sustainability practices, offer valuable insights into building self-sustaining extraterrestrial colonies. By engaging in analog astronaut research, Africa is positioning itself as an essential partner in global space exploration initiatives, ensuring that the continent's knowledge and experience contribute to shaping the future of human life beyond Earth.

Africa's potential contribution to humanity's future beyond Earth is profound [36]—and poetic. The continent that cradled humanity's origins could also cradle its cosmic expansion. Kenya's initiatives hint at a future where Africa leads in asteroid mining (unlocking rare minerals for space habitats), exoplanet research (leveraging its observatories), or even interstellar communication (drawing on its multilingual diversity to craft messages for alien civilizations). From the San people of the Kalahari to the Maasai of East Africa, the Ogiek of Kenya, and the Pygmy groups of Central Africa, diverse African communities have tailored their ecological knowledge to their unique landscapes.

The Maasai, for instance, use their expertise as herders in Kenya and Tanzania to sustainably manage cattle and grazing areas. They track seasonal shifts, practice rotational grazing to avoid depleting the land and live alongside wildlife like elephants and lions in a balanced coexistence. For example, the San people of southern Africa have mastered survival in dry regions [37] with limited water and food, relying on deep ecological knowledge to locate resources like underground water-storing tubers or game trails. In a space context, where water and food are finite, analogous strategies could inspire efficient recycling systems or novel ways to identify and stretch limited supplies. Their use of lightweight, multi-purpose tools—such as the digging stick—also mirrors the need for compact, versatile equipment in space missions.

Africa could become the heart of a multiplanetary civilization that reflects humanity at its best—diverse, adaptive, and interconnected. If the first human footsteps on Mars carry the imprint of African ingenuity, it would close a cosmic circle: the continent where we began as a species shaping where we go next. This is not just about survival beyond Earth; it is about Africa redefining what it means to thrive there, infusing our galactic future with its spirit.

Living in fragile ecosystems—Arctic tundras, high mountains, rainforests, deserts, or coastal zones—Indigenous peoples face climate change head-on. In Asia, they contend with fiercer storms and rising seas; in Africa, with drought and desertification; in the Arctic, with melting ice disrupting hunting. Yet, their deep environmental bond fuels resilience, enabling them to adapt and endure. Their sustainable practices also help counter ecological decline [38].

Remote living often means limited access to modern healthcare [39], exacerbating health disparities. For example, chronic diseases and infectious outbreaks like COVID-19 pose significant risks due to poor infrastructure and distance from medical facilities. Research on rural African communities suggests that Indigenous groups face higher disease burdens [40], with factors like malnutrition and lack of clean water compounding the issue. Yet, their holistic view of health—integrating physical, spiritual, and environmental well-being—offers resilience, as seen in traditional healing practices that persist despite isolation.

Africa's space odyssey is not a footnote—it is a cornerstone. From Kenya's launchpads to the stars, the continent could bridge our past and future, grounding humanity's cosmic ambitions in a story as old as the savannah skies.

Moreover, Africa's participation in space exploration signifies a shift in global perceptions of the continent—from a recipient of technological advancements to a leader in scientific innovation. Through Mars Society Kenya and its OASEAS project, Kenya is setting a precedent for other African nations to invest in space research, fostering continental collaborations that can further strengthen Africa's space presence.

3. Conclusion

Lessons from these parallels might include integrating traditional ecological knowledge into life-support system design, fostering crew resilience through community-building practices, or even training astronauts in observational skills honed by nomadic peoples. Space agencies like NASA or private ventures like SpaceX and Axiom Space could explore collaborations with anthropologists studying these groups to refine analog simulations—say, testing a San-inspired resource foraging protocol in a Mars habitat.

Kenya's strategic location for space activities, with its equatorial spaceport, positions it ideally for satellite launches into geostationary orbits. Although currently underutilized, there is immense potential for Kenya in both commercial and scientific space ventures. The need for educational reform focused on innovations beyond just Low Earth Orbit (LEO) and advocating for training centers could help Kenya become a spacefaring nation by 2030.

Mars Society Kenya's primary mission is to increase public awareness and enthusiasm for space exploration. They aim to inspire Kenyans, particularly the youth, to look beyond the Earth's horizon and consider careers in Science, Technology, Engineering, and Mathematics (STEM). By doing so, they hope to cultivate a generation that can contribute to, and perhaps even lead, future extraterrestrial endeavors.

Building OASEAS and promoting space exploration in Kenya comes with its set of challenges, primarily funding. Space projects require substantial investment, and securing this in a developing economy can be daunting.

Mars Society Kenya envisions a future where Kenya is not just a participant but a leader in space exploration. The Society's efforts are directed toward not only inspiring young Kenyans but also ensuring they have the opportunities to engage in real space research. With projects like OASEAS, they are laying the groundwork for Kenya to be recognized on the global stage in the race to understand and perhaps 1 day settle planets beyond our own.

Mars Society Kenya is more than just an organization; it is a beacon of hope and innovation in the African continent's journey toward the stars. Their work in the Chalbi Desert with OASEAS, their relentless STEM outreach, and their vision for a Kenyan space industry are steps toward a broader horizon where space exploration is accessible to all, regardless of where one comes from. As we watch this Society grow, it becomes clear that the future of space might just have a Kenyan accent.

The bridge here is not just practical; it is philosophical. Indigenous African communities show that survival in extreme conditions is not solely about technology—it is about ingenuity, interdependence, and a deep connection to one's environment. For humans reaching into the void of space, those are roots worth remembering.

Declarations

I, Isaac Macharia, hereby declare that this chapter is an original work, developed through extensive research and collaboration with Mars Society Kenya and various stakeholders in the space exploration community. The contents of this chapter are based on factual information, empirical research, and expert insights into analog astronaut research and its significance in Africa.

This work is intended to contribute to the growing body of knowledge on space science, sustainability, and human adaptation to extreme environments. It reflects my commitment to advancing Africa's role in global space exploration and fostering innovation in space technologies.

All sources referenced have been duly acknowledged, and any errors or omissions are unintentional. I affirm that this chapter has not been published elsewhere and is submitted in good faith for academic and scientific discourse.

Acknowledgements

I would like to extend my heartfelt gratitude to all those who have contributed to the realization of this chapter. My deepest appreciation goes to Mars Society Kenya for their unwavering commitment to advancing space research in Africa and for spearheading the OASEAS habitat project. I am also immensely grateful to the dedicated researchers, engineers, and local communities whose collaboration has been invaluable in shaping this initiative.

Special thanks to the Academic Editor for their insightful guidance and encouragement, which has helped refine the narrative of Africa's role in space exploration. Lastly, I am indebted to my colleagues, friends, and family for their continuous support and belief in the potential of African-led space initiatives. This chapter is a testament to the collective effort to position Africa as a key player in humanity's future beyond Earth.

Conflict of interest

The authors declare no conflict of interest.

Author details

Isaac Macharia
Mars Society Kenya, Thika, Kenya

*Address all correspondence to: igathu@marssociety.org

References

[307] De Vaucouleurs G. Physics of the Planet Mars. Faber and Faber Ltd.; 1954

[308] Kenya Space Agency (KSA). KSA Popular Version. 2017. Available from: https://ksa.go.ke/assets/files/KSA_POPULAR_VERSION_October_Compressed.pdf

[309] Kenya Space Agency (KSA). Reviewed Second Draft Kenya Space Bill; 2024. Available from: https://ksa.go.ke/assets/files/REVIEWED--SECONDDRAFTKENYASPACEBILL2024.pdf

[310] Christian B, Alexander S. Outer Space in Society, Politics and Laws. Washington, D.C.: U.S. Government Printing Office; 2012. pp. 173-180

[311] Mars Society Kenya, Official Website; https://kenyamarssociety.org/about-us/, https://www.facebook.com/100064009360355/posts/its-official-on-to-marsas-of-monday-4th-november-2024-mars-society-kenya-was-off/968008702009433/

[312] Mars Society Kenya, OASEAS Habitat. Available from: https://mdrs.kenyamarssociety.org/chalbi-mission-alpha/

[313] United Nations Permanent Forum on Indigenous Issues (UNPFII). Report on Indigenous Issues. New York, NY: United Nations; 2006

[314] United Nations Department of Economic and Social Affairs (UNDESA). Indigenous Peoples and Sustainable Development. New York, NY: United Nations; 2004

[315] European Space Agency. About Space Debris [Webpage on the Internet]. 2023. Available from: https://www.esa.int/Space_Safety/Space_Debris/About_space_debris [Accessed: April 17, 2024]

[316] International Astronautical Federation. Global Space Exploration Conference (GLEX-2021) (Paper No. 61678). 2021. Available from: https://dl.iafastro.directory/event/GLEX-2021/paper/61678/

[317] Khoza RJ. Let Africa Lead: African Transformational Leadership for 21st Century Business. Johannesburg: Venzutu; 2006

[318] Space exploration and economic growth: New issues and horizons. Luisa Corrado, Maureen Cropper, Akhil Rao. 120(43);e2221341120. DOI: 10.1073/pnas.2221341120

[319] Gopaldas R. The State of the African Space Industry. Singapore: Nanyang Technological University; 2023

[320] United Nations. Declaration on the Rights of Indigenous Peoples (UNDRIP). New York, NY: United Nations; 2007

[321] Christol CQ. 'The International Law of Outer Space', Naval War College Newport, Rhode Island, International Law Studies 1962. Washington D.C.: U.S. Government Printing Office; 1966. pp. 84-88

[322] Library of Congress. Finding Our Place in the Cosmos with Carl Sagan: Seeing and Interpreting Martian Oceans and Canals. n.d. Available from: https://www.loc.gov/collections/finding-our-place-in-the-cosmos-with-carl-sagan/articles-and-essays/life-on-other-worlds/seeing-and-interpreting-martian-oceans-and-canals/

[323] Terblanché-Greece AC. Ubuntu and environmental ethics: The west can learn

from Africa when faced with climate change. In: Chemhuru M, editor. African Environmental Ethics: A Critical Reader. Cham, Switzerland: Springer Nature; 2019. pp. 93-109

[324] Kwasi W. Humanity, African thought. In: Masolo D, editor. African Philosophy. New York, NY: Science JRank Publishing; n.d. Available from: http://science.jrank.org/pages/776/Humanity-African-Thought.html

[325] Agrawal A. Indigenous Knowledge and Resource Management. Cambridge, UK: Cambridge University Press; 2002

[326] Mavhura E, Mushure S. Forest and wildlife resource-conservation efforts based on indigenous knowledge: The case of Nharira community in Chikomba district, Zimbabwe. Forest Policy and Economics. 2019;**105**(May):83-90. DOI: 10.1016/j.forpol.2019.05.019

[327] African Union Commission. Agenda 2063: The Africa We Want Advances in Telemedicine, Disease Models, Psychological Stress Response Systems, Nutrition, Cell Behaviour and Environmental Health are Some of the Benefits Gained from the International Space Station Microgravity Environment; 2015. Available from: https://www.nasa.gov/mission_pages/station/research/benefits/human_health.html

[328] Lee JJ, Chung S. Space policy for late comer countries: A case study of South Korea. Space Policy. 2011;**27**:227-233

[329] Mckinell L. Space science research in Africa. South African National Space Agency (SANSA). In: Presentation at AMSIR Workshop; Boston. 2013. Available from: http://www.bc.edu/content/dam/files/sites/amist/presentations/F0900_Mckin_nell_AMSIR-workshop-Boston.pdf

[330] Ngcofe L, Gottschalk K. The growth of space science in African countries for Earth observation in the 21st century. South African Journal of Science. 2013;**109**(1/2):Article #a001. DOI: 10.1590/sajs.2013/a001

[331] Spencer O, Francis C, Olufemi A, Henry A. Building the appropriate capacity for enabling space programs in Africa: The Nigerian experience. International Journal of Sustainable Development Research. 2019;**5**(1):18-23. DOI: 10.11648/j.ijsdr.20190501.13

[332] Yeshurun Alemayehu Adde (Kibret). Socioeconomic Benefits of Space Technology for Africa, 10:6 IEE-SEM J at 260; 2019

[333] The San Marco Project Research Centre. Centro di Ricerca Progetto San Marco – University of Rome "La Sapienza"; n.d. [Accessed: August 23, 2010]

[334] Lowell P. In Mars and its Canals. New York, NY: Macmillian; 1906

[335] Horowitz NH. The Search for Life on Mars. California Institute of Technology; 1977. Available from: https://calteches.library.caltech.edu/2060/1/horowitz.pdf

[336] Kimberley Declaration. International Indigenous Peoples' Summit on Sustainable Development. Kimberley, South Africa; 2002

[337] Ferreira-Snyman A. Challenges to the prohibition on sovereignty in outer space: A new frontier for space governance. Potchefstroom Electronic Law Journal. 2021;**24**:1-50

[338] Redfield P. The half-life of empire in outer space. Social Studies of Science. 2002;**32**:791-824

[339] United Nations Institute for Disarmament Research. The Role of

Norms of Behaviour in African Outer Space Activities [Document on the Internet]. 2012. Available from: https://unidir.org/files/publication/pdfs/the-role-of-norms-of-behaviour-in-african-outer-space-activities-en-418.pdf [Accessed: October 10, 2024]

[340] Sacred Natural Sites: Conserving Nature and Culture. Available from: https://portals.iucn.org/library/sites/library/files/documents/2010-045.pdf

[341] Illich I. Tools for Conviviality. New York: Harper and Row; 1973

[342] Vakoch DA, Harrison AA, editors. Civilizations beyond Earth: Extraterrestrial Life and Society. New York, NY, UK, USA, Oxford: Berghahn Books; 2011

[343] United Nations Office for Outer Space Affairs. Space for Agriculture Development and Food Security: Use of Space Technology within the United Nations System. Vienna, Austria: United Nations Office for Outer Space Affairs; 2016. p. 5. Available from: https://www.unoosa.org/res/oosadoc/data/documents/2016/stspace/stspace69_0_html/st_space_69E.pdf

[344] Salick J, Byg A. Indigenous Peoples and Climate Change. Copenhagen, Denmark: International Work Group for Indigenous Affairs; 2007

[345] Mbembe A. Pathways of tomorrow: Contribution to thinking commensurate with the planet. In: Education Research and Foresight Working Paper 32. Paris: UNESCO; 2023

[346] Pyatibratov GY, Kravchenko OA, Kivo AM. Design principles and implementation of advanced simulators for training astronauts to work in zero or low gravity conditions. Procedia Engineering. 2016;**150**:1410-1414. DOI: 10.1016/j.proeng.2016.07.337

Chapter 10

Perspective Chapter: Beyond the Algorithm – The Irreplaceable Role of Human Doctors in Space

Bashar Othman, Sara Rothrock and Peter H.U. Lee

Abstract

As human spaceflight moves from the relative safety of low Earth orbit to the Moon, Mars, and beyond, the demands on medical care will change in unprecedented ways. This chapter examines the evolving role of the "space surgeon," a dedicated, highly trained physician prepared to address the full spectrum of medical care in an environment where evacuation is improbable, resources are constrained, and communication delays hinder real-time guidance from Earth. Drawing on lessons from six decades of human spaceflight, the unique physiological, operational, and psychological challenges of long-duration missions are discussed, including microgravity-induced physiological impairments, radiation exposure, and the stresses of isolation and confinement. This chapter evaluates how emerging technologies like autonomous diagnostics, robotics, and artificial intelligence, can extend medical capabilities in space, while underscoring their limits in replacing human judgment, adaptability, and compassion. Through an integrated discussion of risks, operational constraints, and mission-critical contingencies, an argument is made that despite rapid advances in medical technology, the presence of a skilled human physician will remain indispensable for safeguarding crew health, ensuring mission resilience, and enabling the next era of deep space exploration.

Keywords: space medicine, space surgery, space surgeon, long-duration spaceflight, autonomous medical systems

1. Introduction

After more than half a century of continuous human spaceflight, the United States is preparing to take its next major step outward. The Artemis program aims to return humans to the lunar surface, this time with the goal of sustained operations and infrastructure. Beyond the Moon, mission planners are actively considering the enormous technical, logistical, and human challenges of sending a crew to Mars. These ventures represent a paradigm shift from the last 60+ years of primarily low Earth orbit (LEO) operations.

Human spaceflight has so far remained almost entirely within LEO, where proximity to Earth has allowed most health-related risks to be effectively managed [1]. In contrast, exploration-class missions to the Moon and Mars involve fundamentally different operational constraints due to their substantial distance from Earth, requiring new mission architectures and improved onboard medical systems. The five health-related hazards of spaceflight—space radiation, altered gravitational forces, isolation and confinement, hostile closed environments, and physical separation from Earth—are all significantly intensified in deep space missions due to increased mission durations and round-trip transit times [1]. Collectively, these hazards drive multisystem physiological changes that heighten the risk of significant in-flight medical events during exploration-class missions relative to LEO. The principal hazards and their associated physiologic effects are depicted in **Figure 1** [2].

In these missions, evacuation to Earth will be unrealistic, resupply capabilities will be unlikely, and strict mass and volume constraints will limit medical infrastructure. Adding to the challenge, communication delays of several minutes each way will complicate clinical decision-making and operational coordination [3]. These realities demand a rethinking of not only spacecraft systems and crew composition but also the fundamental approach to medical care in space.

In current space operations, astronaut health is supported primarily by flight surgeons—Earth-based physicians responsible for pre-flight readiness, in-mission monitoring, and post-flight rehabilitation. Onboard, medical care typically falls to

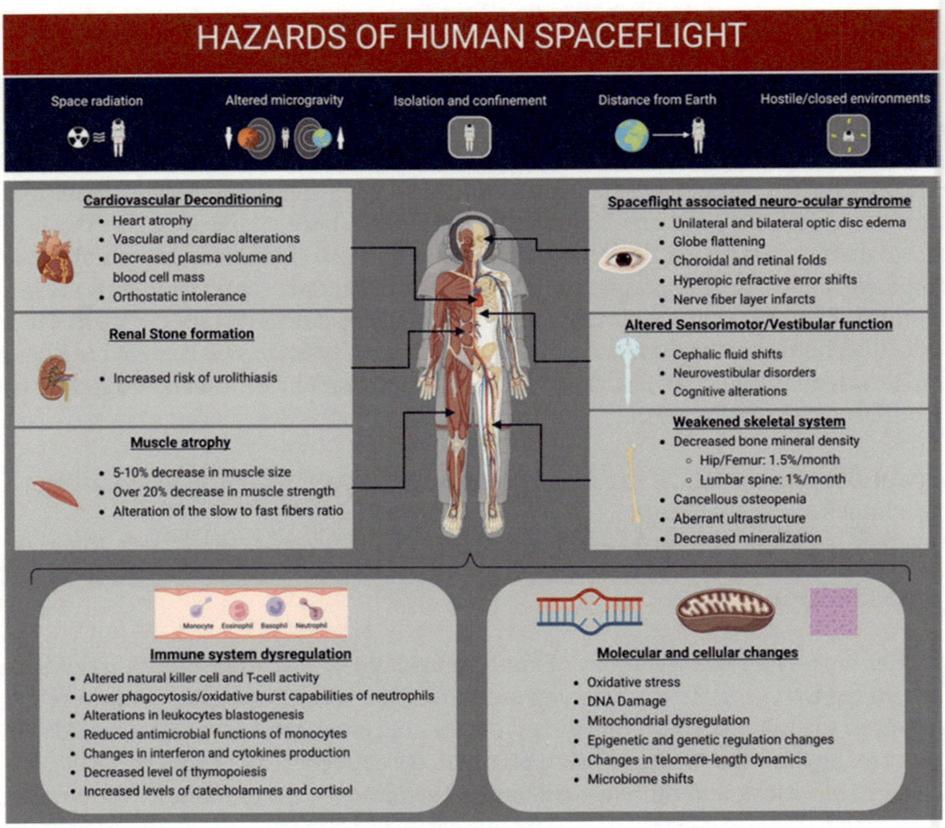

Figure 1.
Principal hazards of human spaceflight and associated health effects. The five core hazards of crewed missions—space radiation, altered gravity, isolation and confinement, distance from Earth, and hostile closed environments—drive multisystem physiologic changes that can compromise crew health, degrade performance, and jeopardize mission success. Reference: Camera et al. [2].

the Crew Medical Officer (CMO), often a non-physician astronaut with supplemental medical training. When a physician has flown, their contributions have generally been limited to minor procedures and acute care management. While no mission has yet required major surgery, this fortunate record reflects careful crew selection, proper risk mitigation, and some luck rather than the inherent safety of the environment. Flight surgeons who happen to be astronauts are not specifically trained for autonomous, complex surgical interventions in space, and CMOs without medical degrees face even greater constraints—limitations that become critical in exploration-class missions where Earth-based backup will be delayed, limited, or unavailable. Addressing these challenges has led to the concept of a "space surgeon": a dedicated medical officer trained to manage the full spectrum of health concerns, from primary care to surgical emergencies, entirely independent of Earth-based intervention [4]. Beyond treating illness and injury, the space surgeon would also be a central figure in preventive health, performance optimization, and management of mental health.

In this chapter, we examine the physiological and operational medical challenges unique to long-duration missions, review the evolving role of medical personnel in spaceflight, assess the capabilities and limitations of emerging technologies, identify persistent gaps in care, and argue why human medical expertise remains essential and irreplaceable.

2. Unique medical challenges in long-duration spaceflight

2.1 The hazards of human spaceflight

The space environment is hazardous for humans, and includes altered gravity, space radiation, isolation and confinement, hostile closed environments, and distance from Earth. Its unique physical and operational conditions not only increase the risk of medical problems but also fundamentally alter how health is maintained, how illnesses manifest, and how treatments can be delivered. Long-duration missions, such as those to the Moon or Mars, will magnify these challenges and strain the boundaries of current space medical capabilities.

Microgravity remains the most defining feature of space. In the continuous free-fall environment of orbit, almost every physiological system in the human body undergoes some degree of adaptation. These changes affect how astronauts live, work, and move, but they also alter every step of medical care—from basic procedures to complex surgeries. Blood, surgical fluids, and even the position of organs behave differently, complicating both diagnosis and treatment.

Space radiation represents a more insidious threat. Invisible to human senses and without the benefit of Earth's thick atmosphere or magnetic shielding, crews in deep space will be subjected to significantly elevated cosmic and solar radiation levels. Outside the Van Allen belts, the risk of developing serious medical conditions rises substantially. Almost all of our operational experience in human spaceflight has been in LEO, leaving our understanding of deep space radiation effects—and our mitigation strategies—limited.

Isolation and confinement add another layer of complexity [5]. Psychological strain is one aspect, but from a medical standpoint, the inability to summon immediate help or evacuate a critically ill crewmember is far more consequential. On a Mars mission, communication delays can reach more than 40 minutes round-trip, making real-time telemedicine impossible and demanding medical autonomy at the crew level.

Human spaceflight relies on a closed, actively controlled habitat. Environmental Control and Life Support Systems (ECLSS) regulate cabin pressure, O_2/CO_2, temperature, humidity, water and air quality, and acoustics. Even under nominal conditions, elevated CO_2, low humidity, noise and vibration, and trace contaminants can impair mucosal defenses, cognition, sleep, and cardiorespiratory function. Off-nominal excursions or system malfunctions—e.g., CO_2 spikes, water-system contamination, or an ammonia leak—can quickly become medical threats, underscoring the need for robust monitoring and onboard mitigation.

Physical separation from Earth is itself a medical risk factor: after trans-Mars injection there is no rapid abort, evacuation, or resupply, and every emergency must be managed on board with finite consumables and capability. One-way light-time delays of up to ~22 minutes on a Mars mission preclude real-time telemedicine, relegating flight surgeons to asynchronous consultants while the crew relies on mission-specific training, offline decision support, and highly reliable, repairable systems. As communication latency grows, real-time telemedicine and ground support progressively erode. **Figure 2** maps this shift, showing how time-critical problem solving, clinical decision-making, and even vehicle control move to the crew as delay increases, with only non-urgent tasks suited to delayed Earth-based input [6].

2.2 Physiological effects of spaceflight

More than 60 years of human spaceflight have yielded valuable insights into how the body adapts to microgravity. These changes begin within seconds of orbital insertion and evolve toward a new physiological setpoint, or "deep adaptation," over 2 to 6 weeks [7]. While the human body can adjust, these adaptations carry risks for both acute and chronic health conditions—and in deep space, prolonged exposure may reveal new hazards not yet seen.

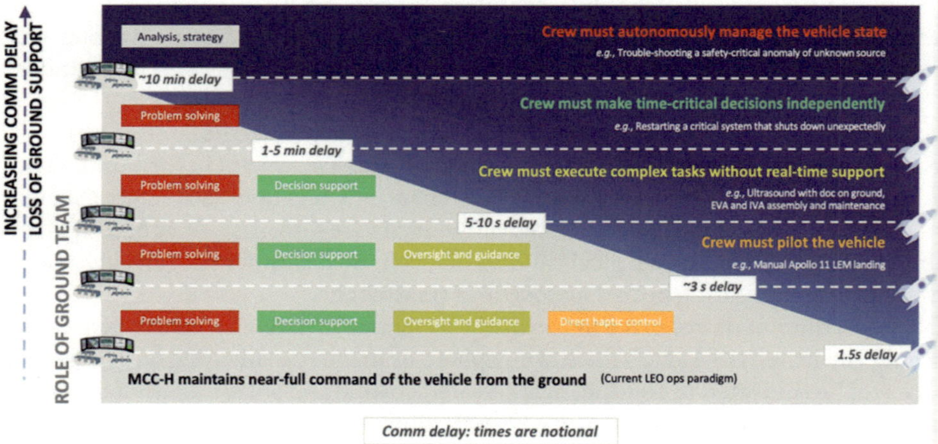

Figure 2.
Progressive ground-to-crew shift in operational and medical responsibility with increasing communication delay. As Earth-spacecraft one-way light-time increases, the ground team's role declines from direct control and real-time guidance to asynchronous decision support, while the crew assumes piloting, complex task execution without real-time support, and time-critical independent decision-making—including autonomous medical care when real-time telemedicine is not possible. Only non-urgent or strategic issues remain suited to delayed Earth input. Abbreviations: MCC-H, Mission Control Center-Houston; EVA, extravehicular activity; IVA, intravehicular activity; LEM, Lunar Excursion Module; LEO, low Earth orbit. Reference: Landon et al. [6] https://ntrs.nasa.gov/api/citations/20250003885/downloads/NASA%20TM20250003885.pdf.

The musculoskeletal system is among the most visibly affected. Muscle atrophy can be detected within 2 weeks of weightlessness, with postural muscles estimated to lose up to 10% of their mass [8]. This loss reduces strength, mobility, and the ability to perform critical mission tasks—such as self-extrication after a hard planetary landing. Bone demineralization begins just as quickly, increasing fracture risk and contributing to renal stone formation due to altered calcium metabolism [9, 10].

The cardiovascular system also undergoes marked changes. Cephalad fluid shifts occur almost immediately, producing nasal congestion, facial puffiness, and a perceived fluid overload [11]. This shift contributes to space motion sickness (SMS), which affects roughly 60–80% of astronauts in the first days of flight [12]. Over time, reduced plasma volume, cardiac atrophy, and decreased red blood cell mass impair orthostatic tolerance, creating challenges for planetary landings where ground medical support is absent [13, 14].

Gastrointestinal and renal functions adapt in less obvious but operationally important ways. Slowed bowel motility and urinary retention are common, and renal stones are already a concern in microgravity—are particularly challenging to diagnose and treat in-flight [15, 16].

The immune system also changes in flight. Both innate and adaptive immunity may be suppressed, leading to reactivation of latent viruses, increased infection susceptibility, and slower wound healing [17]. Combined with the altered behavior of microbes in microgravity, infection control becomes a heightened concern.

Visual and neurological changes have emerged as significant long-duration concerns. Spaceflight-associated neuro-ocular syndrome (SANS), likely linked to fluid shifts impairing cerebral venous drainage, can cause decreased visual acuity, optic disc edema, and structural changes to the globe [18, 19]. Neurovestibular disturbances—spatial disorientation, motion control impairment, and balance deficits—are common both in the early days of flight and upon return to gravity, affecting operational safety during EVA or planetary surface activity [20].

Space radiation remains an unavoidable hazard in deep space. Galactic cosmic rays (GCRs) provide a chronic background dose, while solar particle events (SPEs) can deliver intense acute exposures [21]. Even with shielding strategies, a 950-day Mars mission could deliver cumulative doses exceeding 450 mGy, especially during the ~450 days of transit [7]. These exposures may heighten cancer, cardiovascular, and central nervous system risks, potentially interacting with microgravity and isolation to accelerate cognitive or psychiatric decline [21, 22].

Finally, psychological and cognitive stressors are pervasive. Prolonged isolation, confinement, circadian rhythm disruption, and separation from Earth can degrade mental health and team cohesion [23]. Cognitive performance may be impaired both by these stressors and by direct radiation effects [22]. While current countermeasures—psychological training, private communication with family, and crew compatibility screening—are useful, they may be insufficient for multi-year missions to Mars or beyond [24].

The medical risks of long-duration spaceflight are multifactorial, cumulative, and often synergistic. Musculoskeletal and cardiovascular deconditioning, immune dysregulation, neuro-ocular changes, radiation exposure, and behavioral health stressors all threaten crew safety and mission success. Each astronaut will respond differently, adding a layer of unpredictability to prevention and treatment strategies. The limitations of in-flight diagnostics and therapeutics only compound these risks.

2.3 Operational constraints in isolated environments

Beyond the physiological impacts of spaceflight, the operational realities of providing medical care in deep space present a distinct set of challenges. Exploration-class missions will require crews to operate with complete medical autonomy. Evacuation to Earth and resupply capabilities will be unrealistic and improbable. Any injury, illness, or surgical emergency will need to be managed with the personnel, training, and equipment already aboard the spacecraft. This constraint elevates the importance of preventive medicine, robust contingency planning, and versatile crew skillsets.

Resource limitations will be a constant factor. Medical supplies, diagnostic devices, and therapeutic tools must be carefully selected for maximum utility while adhering to strict mass, volume, and power budgets. Consumables such as oxygen, intravenous fluids, or medications cannot be replenished in-flight, necessitating careful rationing and planning for reuse or regeneration wherever possible. Equipment redundancy will be limited, and repair in microgravity can present its own hazards. Moreover, surgical and diagnostic equipment are often bulky, heavy, and designed for terrestrial environments, making them impractical for inclusion on most missions. Even when available, their utility may be limited to rare or highly specific scenarios. As a result, the vast majority of conventional surgical instruments, imaging systems, and specialized devices will not be carried, significantly restricting the range of procedures that can be performed. This reality demands the use of alternative modalities, multipurpose tools, and novel strategies tailored for austere, resource-limited space environments.

Zero-gravity working conditions impose unique procedural challenges. Simple tasks like wound cleaning, suturing, or even positioning a patient require adaptations to prevent contamination and control fluids. For more invasive procedures—such as chest tube insertion, fracture reduction, or open surgical procedures—additional constraints arise from limited anesthesia capability, restricted operative lighting, and the challenges with maintaining a fully sterile field and containing bodily fluids.

Communication delay is another operational reality for lunar and especially Martian missions. While lunar crews will experience only a few seconds' delay in voice and data exchange, Mars-bound astronauts will contend with one-way delays of up to 22 minutes, depending on planetary alignment—up to 44 minutes round trip. This latency makes real-time consultation with ground-based flight surgeons impossible, shifting decision-making authority entirely to the onboard crew. Artificial Intelligence (AI) and autonomous decision-support systems may help fill this gap, but they cannot replace the situational awareness, adaptability, and ethical judgment of a human medical professional in the moment.

Historical in-flight medical interventions offer useful lessons. For example, the extraction of a loose foreign body from an astronaut's eye on the International Space Station (ISS), the use of ultrasound-guided imaging for suspected deep vein thrombosis, and the improvised dental repairs performed on Mir demonstrate the feasibility of certain procedures in microgravity. Each case also highlights the constraints of limited equipment, small medical inventories, and the need for ingenuity when facing unexpected problems. These successes underscore the value of thorough pre-flight training and multidisciplinary skill development—not just for medical officers, but for all crew members.

In essence, medical care in deep space is defined by autonomy, adaptation, and austerity [25]. The absence of immediate rescue or resupply means that the margin for error is small and the stakes are high. The space surgeon, ideally with advanced procedural skills and broad clinical experience, must be prepared to handle any scenario—from minor ailments to life-threatening emergencies—under conditions far removed from the controlled, resource-rich environments of Earth-based medicine.

3. Evolution of the space surgeon

3.1 Spectrum of medical care

Physicians, whether on Earth or in orbit, serve across a broad continuum of patient care. This spectrum of medical care encompasses preventive medicine, health surveillance, diagnosis, treatment (both medical and surgical), and rehabilitation. In the terrestrial setting, these phases of care occur within an extensive infrastructure of hospitals, laboratories, pharmacies, rehabilitation centers, and specialist networks. In space, that infrastructure must be distilled into a compact, self-sufficient capability—and the space surgeon must be prepared to operate across the entire continuum, often with limited resources, environmental hazards, and no possibility of evacuation. The five core domains of the spectrum of care, as they apply to spaceflight, are outlined below.

Preventive medicine: Preventive care in space begins long before launch. It encompasses crew selection, preflight screening, immunizations, optimization of chronic health conditions, and physical conditioning [26]. The aim is to minimize the likelihood of in-flight medical events by identifying and mitigating risk factors in advance. In-flight preventive measures focus on counteracting known hazards of the space environment. These include exercise countermeasures, tailored nutrition, hygiene protocols, environmental monitoring, and prophylactic medications. Preventive medicine also extends to behavioral health, ensuring that psychological support systems and morale-maintaining activities are in place to reduce stress and interpersonal conflict [5].

Surveillance: Medical surveillance involves the continuous or periodic monitoring of crew health status. In space, this includes both physiological measurements (e.g., vital signs, cardiac rhythm, bone density, visual acuity) and environmental monitoring (e.g., air quality, radiation dose, microbial contamination). Surveillance may be largely delegated to trained non-physician crew members, provided they have clear protocols for when to escalate findings to the space surgeon [27]. Automated monitoring devices and wearable sensors can support this process, but interpretation of subtle patterns and trends—such as early signs of immune dysregulation or neuro-ocular changes—still benefits from physician oversight [28]. Timely recognition of a deteriorating trend can allow intervention before a minor issue becomes a mission-threatening emergency.

Diagnosis: Accurate diagnosis in space is complicated by limited diagnostic tools, environmental influences on physiology, and the possibility of unfamiliar disease presentations. Medical personnel are often forced to work with incomplete information, integrating crew-reported symptoms, physical exam findings, and available laboratory or imaging data [28]. While certain protocols can guide non-physician crew through basic diagnostic steps, more complex or ambiguous presentations require the physician's ability to synthesize disparate clues, consider alternative explanations, and recognize rare but high-impact conditions. This is particularly important when deciding whether to initiate resource-intensive treatments in an environment where supplies are finite.

Medical and surgical treatment: Treatment in space encompasses a wide range of interventions, from administering medications for common ailments to performing emergency surgery under microgravity or reduced gravity conditions [28, 29]. Some medical treatments can be carried out by any crew member following detailed protocols—for example, managing mild motion sickness, rehydrating after exercise, or applying a splint to a minor injury [30]. However, emergent care presents unique challenges. Severe trauma, hemorrhage, airway compromise, or acute surgical emergencies (e.g., appendicitis, cholecystitis) require rapid, skilled intervention. In such cases, a space surgeon must act without delay, often without the ability to consult in real-time with Earth. Even when telemedicine is possible, communication delays—particularly during Mars missions—make immediate on-site expertise critical. Surgical care in space is further complicated by microgravity, limited sterilization capability, constrained equipment, and the need to prevent contamination of the cabin environment [31].

Rehabilitation: Rehabilitation in space focuses on restoring function after injury, illness, or prolonged microgravity exposure [32]. This may include physical therapy to recover muscle strength and joint mobility, cardiovascular conditioning, and psychological support after traumatic events [32]. On long-duration missions, rehabilitation is essential not only for returning to pre-injury status but also for enabling crew members to continue contributing to the mission. This might involve adapting rehabilitation techniques to the spacecraft environment, using resistive exercise devices or modified mobility aids, and integrating recovery programs into the daily mission schedule without overtaxing limited resources.

The spectrum of medical care in space mirrors that on Earth in its stages but differs profoundly in its execution. While technology and cross-trained crew can assume some responsibilities, the presence of a space surgeon will ensure that the most complex, urgent, and high-risk aspects of care are managed with the full breadth of human clinical capability. In exploration-class missions, where medical autonomy is absolute and the margin for error is vanishingly small, this breadth of expertise may be the difference between mission success and mission failure.

3.2 Flight surgeons: Guardians from the ground

Since the earliest days, flight surgeons have been the primary doctors for astronauts, though they traditionally remain on Earth. NASA flight surgeons are physicians specialized in aerospace medicine are often assigned to specific spaceflight missions. They begin working with the astronaut crew months to years before launch, overseeing their health, conditioning, and preparation [33]. They continue support during the mission from Mission Control and follow up with astronauts post-mission through rehabilitation. A flight surgeon's duties are multifaceted: pre-mission, they ensure the crew meet rigorous medical standards and help devise individualized countermeasure plans (exercise, diet, etc.) [33]. In flight, they sit on console in the Mission Control Center, monitoring incoming health data and advising the crew [27]. They conduct private medical conferences with the crew, addressing issues from minor symptoms to mental health check-ins [27]. They are also the coordinators in case of a medical emergency, calling on consulting specialists (cardiologists, surgeons, etc.) when needed [27]. In short, a flight surgeon acts as the personal physician for astronauts, albeit via remote care.

Becoming a NASA flight surgeon requires extensive training. Typically, these individuals are medical doctors who undergo residency in aerospace medicine (often after another specialty like family medicine or emergency medicine). They learn about hyperbaric medicine, radiation health, the physiology of extreme environments, and space-specific medical hardware. Continuous education is necessary to keep up with new spaceflight medical protocols. NASA also cross-trains flight surgeons internationally; for ISS, a U.S. flight surgeon often works alongside a Russian counterpart, and they must understand each other's systems and approaches [33]. Anecdotally, flight surgeons often describe their role as equal parts doctor, engineer, and "crew family member"—they must build trust with astronauts such that an astronaut will report a headache or anxiety episode in space without fear of it being misjudged. This close relationship is crucial, as early intervention can prevent small issues from becoming mission-threatening crises.

3.3 The evolution of the "space surgeon": A doctor in the crew

In practice, missions to date have not carried a doctor on every crew. Instead, one astronaut per crew increment is designated the CMO after undergoing extra medical training which may include intensive courses in emergency medicine and paramedic skills [27]. However, for missions beyond low Earth orbit, it will likely be necessary to including a highly trained medical officer onboard who is a physician. This role is informally dubbed the "space surgeon"—essentially an astronaut who, in addition to other duties, serves as the "ship's doctor" and is capable of performing surgical and emergency medical procedures in space [4]. Unlike a flight surgeon, who stays on Earth, a space surgeon would live and work alongside the crew. The need for this role stems from the recognition that, for the Moon, Mars, and deep-space voyages, evacuation is impractical and communication is delayed, so at least one crew member should be able to act as a physician-surgeon on the spot. The space surgeon would also oversee routine health maintenance for the crew, conducting periodic exams and guiding exercise and nutrition plans [29]. In essence, the space surgeon is the extension of the flight surgeon on the ground, but with the added ability to provide hands-on treatment immediately. Their presence is seen as critical for missions where waiting hours (or days) for help is not an option [28].

NASA has been developing medical checklists and procedural guides that a space surgeon or any crew medical officer can use [30]. These are written in plain language and heavily illustrated so that in an emergency, crewmates could even help talk a non-physician through a procedure [30]. There has been discussion of equipping future spacecraft with augmented reality (AR) glasses that a caregiver could wear to see step-by-step holographic instructions overlapped on their field of view (for example, highlighting exactly where to make an incision). This is an active area of research and could dramatically increase the capabilities of crews without a specialist.

While "space surgeon" is not yet an official NASA title, the concept underscores a philosophical shift: human medical expertise must travel with the crew as we venture farther. Ground support will always be limited by distance [28]. The success of a years-long mission could depend on the decisions of one medically skilled astronaut in a critical moment. Therefore, selecting and training such individuals is a top priority. Just as early space missions picked the best test pilots, future missions may prioritize crews with medical and technical hybrid skill sets—for example, a geologist who is also a trauma surgeon, or an engineer who is also a paramedic [28]. The ultimate expression of this might be the crew of a Mars mission composed so that between them, every major skill (medicine, engineering, navigation, etc.) is covered by at least one person in-depth [28, 29].

This fundamental change to in-flight medicine necessitates new approaches to the selection and training of space physicians, requiring a unique combination of skills not encompassed within any existing training paradigm. These competencies draw from aerospace medicine, emergency medicine, family medicine, and surgical specialties, providing the breadth needed for autonomous care in a highly unpredictable environment.

To complement existing systems, the space surgeon would not replace the flight surgeon but instead assume immediate clinical authority onboard, with Earth-based teams providing longitudinal oversight and asynchronous consultation. When communication windows allow, subspecialists could still contribute guidance, but final responsibility would rest with the onboard physician. Introducing this role fundamentally alters the current astronaut health model, shifting authority from Earth to the spacecraft. Such a position would likely be required on both governmental and commercial missions, as the underlying medical risks remain essentially the same.

Embedding a physician within the crew may also reshape astronaut culture, fostering earlier symptom reporting, strengthening trust in medical oversight, and reinforcing the expectation that health and safety must be managed at the point of risk rather than deferred to ground control. Emerging technologies—including AI-enabled diagnostics, augmented reality for procedures, and continuous wearable monitoring—could further personalize care, ensuring that the space surgeon's expertise is directed to the individual needs of each crewmember, even under conditions of isolation and delay. See **Table 1** for a comparison of current versus future space mission medical practices.

3.4 Procedures and surgical care in long-duration spaceflight

Future exploration-class missions will inevitably encounter medical problems that require invasive diagnostic or therapeutic procedures, despite the best efforts at prevention. These may range from relatively simple interventions—such as intravenous

Current ISS practice	Future exploration missions
CMO role: The crew medical officer (CMO) is often a nonphysician astronaut with abbreviated, protocol-driven medical training.	*Space surgeon role*: A fully trained physician (often with specialty/fellowship expertise) designated to provide comprehensive medical care.
Ground communication: Continuous, low-latency communication with flight surgeons on Earth.	*Delay*: One-way communication delays of ≈3–22 minutes (Mars), limiting or precluding real-time support.
Decision support: Medical decisions are supported and verified in real time by mission control.	*Autonomy*: The space surgeon must make independent clinical and ethical decisions, with only asynchronous consultation available.
Evacuation: Contingency return to terrestrial hospitals may be feasible within hours.	*No evacuation*: Evacuation is impossible; all emergencies must be managed on board for the mission duration.
Behavioral health: Psychological support readily available via frequent Earth contact.	*Behavioral health*: Isolation and delay amplify psychological and ethical stressors; support is provided by trained crew with asynchronous reach-back.
Training model: Condensed, foundational training with reliance on real-time ground verification.	*Training model*: Mission-specific, advanced training focused on autonomous, high-quality care without ground support.

Table 1.
Comparison of medical practice in current vs. future space missions.

line placement, blood draws, or the suturing of minor wounds—to more advanced procedures, including chest tube insertion, central venous access, or percutaneous drain placement. At the far end of the spectrum are surgical procedures. Minimally invasive surgical techniques, such as laparoscopy, are unlikely to be practical due to the mass and volume of required equipment, the specialized skills involved, and the low probability of their necessity [29]. However, open surgical procedures may become unavoidable under certain conditions. Performing such interventions in deep space presents unique challenges—not only in securing the necessary instruments, equipment, and consumable supplies, but also in ensuring that the crew has the required skill sets [29].

On Earth, surgical care typically involves multiple trained professionals, including a surgeon, anesthesiologist, surgical technologist, and circulating nurse. In a spacecraft, these roles would likely have to be assumed by the space surgeon, assisted only by a non-medical crewmate trained to serve as a procedural assistant. This demands a broad and deep skill set that extends well beyond the scope of any single current medical specialty. While most flight surgeons today are dual-trained in aerospace medicine and fields such as emergency medicine, family medicine, or internal medicine, they may also come from surgical disciplines. Yet no existing training program fully prepares a physician for the breadth of care required in space, which would need to encompass elements of surgery, anesthesia, emergency medicine, critical care, behavioral health, and operational medicine—all adapted for an austere, resource-limited environment with little or no outside support.

Most procedures—especially invasive ones—will likely require direct human intervention. However, not all will require a physician, and some could be performed by non-physician crew members given sufficient training and the right equipment. Determining whether a space mission's medical officer must be a physician or surgeon will depend on the anticipated range of medical conditions, the level of additional training provided to non-physician crew, and the scope of available medical equipment. For deep-space missions, however, the more capable and versatile the onboard medical professional, the greater the odds of successful management when medical contingencies arise far from Earth.

4. Technologies that augment medical care

To address the medical risks of long-duration spaceflight, recent efforts have focused on integrating advanced technologies—autonomous diagnostic platforms, robotics, AI, and telemedicine. These tools have demonstrated promise in augmenting clinical decision-making and extending medical capability in resource-limited environments. In certain domains, AI has already demonstrated the potential to outperform humans in diagnostic accuracy, image interpretation, and decision support.

4.1 Artificial intelligence

Crewed deep space missions will demand unprecedented levels of medical autonomy. In these settings, communication delays of up to 44 minutes round trip, absence of evacuation capability, and severe constraints on medical resources will fundamentally alter clinical workflows [34, 35]. AI systems can help bridge the gap between limited onboard human expertise and the broad spectrum of possible medical contingencies [34, 35].

An effective AI medical assistant for spaceflight must function entirely offline, integrate comprehensive medical databases, and conduct symptom-driven diagnostic dialogs [34, 35]. It should obtain targeted histories, guide physical examination and relevant testing, and generate management plans adapted to the available resources [34]. For acute presentations such as chest pain, the system must identify potentially life-threatening conditions—myocardial infarction, pulmonary embolism, pneumothorax—and initiate emergency protocols before a definitive diagnosis is reached [34]. In addition, it should offer just-in-time procedural training, such as echocardiography guidance, to refresh skills during critical moments [34].

AI applications extend across multiple mission-critical health domains. In behavioral health, conversational agents with embedded affect recognition can monitor mood, detect early signs of depression or cognitive decline, and deliver evidence-based coping strategies [35]. For radiation hazards, machine learning models can support early detection of acute radiation sickness through automated cytogenetic analysis and personalized dose-response predictions [35]. Diagnostic imaging interpretation, including ocular ultrasound for SANS, can be enhanced by AI algorithms trained for rapid abnormality detection [21]. Predictive analytics for infection and sepsis, using continuous vital sign monitoring, can prompt early intervention and optimize fluid or vasopressor therapy [35]. Cardiovascular deconditioning risk can be mitigated with closed-loop systems that detect hypovolemia and adjust countermeasures in real time [35].

Operational reliability is essential. The system must run continuously, process streaming physiological data from wearable sensors, and respond instantly to medical queries ranging from drug dosage intervals to equipment troubleshooting [34]. Voice-activated, hands-free operation is particularly advantageous when the crew medical officer is engaged in procedures [34]. The AI should also integrate pre- and inflight medical records with a detailed inventory of onboard consumables to propose feasible treatment pathways and forecast supply usage for the remainder of the mission [34, 35].

While AI can augment diagnostic accuracy, training retention, and procedural guidance, it remains an adjunct rather than a replacement for a trained medical professional [34, 35]. The adaptability, ethical judgment, and improvisational skills of a human physician are indispensable in scenarios where incomplete data, environmental constraints, and unexpected events are a prominent concern. Likely, the optimal approach will pair human expertise with AI-enabled decision support to maximize safety and clinical capability in remote and austere environments.

4.2 Telemedicine and robotic surgery

Telemedicine refers to the remote delivery of clinical services through communication and data technologies. In spaceflight, it has become an essential component of medical care. Telemedicine is typically implemented in one of three formats: synchronous (real-time audio and video), asynchronous (store-and-forward of medical data), and remote monitoring (direct video monitoring or continuous monitoring through tests collected remotely) [7, 36]. On the ISS, synchronous telemedicine enables mission control physicians to guide crew through procedures such as ultrasound exams or minor surgical interventions [36]. However, synchronous communication is only viable for near-Earth missions due to communication latency. For future exploration-class missions, asynchronous methods and increased onboard autonomy will be essential [7]. AI-driven clinical decision support systems and expert knowledge bases embedded in onboard computers may allow crew to perform basic triage, recognize high-risk patterns, and execute emergency protocols without real-time support [3]. In these circumstances, the effectiveness of telemedicine tools will depend heavily on well-designed interfaces, crew familiarity, and the availability of just-in-time procedural training [7]. The ability to interpret nuanced findings or manage complications will likely continue to rely on clinical reasoning supported by pre-mission training.

Telemedicine can also help compensate for the limited medical training of the crew, especially for non-physician CMOs. Telemedicine enables access to the broader range of medical expertise required for complex diagnostic reasoning, procedural guidance, and interpretation of imaging or laboratory data [36]. In communication-limited scenarios, these technologies must be complemented by decision-support tools and just-in-time training modules to ensure adequate care delivery.

Proposals to incorporate robotic surgery in space have garnered attention but remain impractical in the near term. Sophisticated surgical robotic systems like the Da Vinci Xi (Intuitive Surgical, Sunnyvale, CA) are large, heavy, and power-intensive, and their utility in a microgravity spacecraft is unproven. Furthermore, if intended for remote operation, communication delays—particularly for Mars missions—would render real-time control impossible. Previous studies were conducted that simulated communication delays in five test procedures. The latency was changed between 100 ms and 2 s. The performance of the astronauts was significantly degraded with larger latency [37, 38]. Based on these studies, current telesurgery methods are not feasible for exploration-class missions, as there are longer latency periods, which may lead to major surgical failure. However, surgical robots may be advantageous in the future of spaceflight. Robotic surgery has been proven to extend the dexterity and capability of a surgeon in many complex procedures. Some examples of promising technologies include the "Trauma Pod," the RAVEN (Bio Robotics Lab, University of Washington, Seattle WA), or the Robonaut2 (NASA Johnson Space Center) [39, 40].

4.3 Diagnostics and imaging capabilities

Diagnostic capabilities in space are centered around ultrasound. It is the only multipurpose imaging modality with well-established use in microgravity and is deployed for evaluation of abdominal, musculoskeletal, vascular, ocular, and cardiac structures. Ultrasound devices on the ISS support standard imaging modes (B-mode, M-mode, and Doppler) and can be operated by non-expert users under real-time guidance from Earth-based specialists [7, 41]. This has allowed successful remote diagnosis of a range of conditions including venous thrombosis and musculoskeletal injuries [7]. For long-duration missions, where communication delays will preclude synchronous support, the ability to acquire and interpret quality imaging autonomously will be critical (**Figure 3**).

Ultrasound's suitability in space is due to its compact design, low power requirements, and absence of ionizing radiation. Its major limitation is the operator-dependent nature of image acquisition. To address this, ground-based radiologists participate in "tele-guided" scans, using video and voice communication to guide probe placement and optimize imaging planes. Also, just-in-time training methods are employed to familiarize crew members with the device and enhance image quality when they serve as operators [42]. Video latency from the ISS is minimal and does not hinder real-time procedural support [43]. However, for Mars missions with communication delays of up to 46 minutes, tele-guidance becomes impractical. Under such

Figure 3.
Ultrasound aboard the International Space Station. In the Columbus laboratory, NASA astronaut Chris Cassidy performs an ultrasound examination on ESA astronaut Luca Parmitano. Ultrasound's compact size, low power consumption, and broad diagnostic utility make it well suited for spaceflight. Image credit: NASA.

constraints, diagnostic success will depend on crew proficiency and the availability of preloaded procedural guidance or autonomous imaging software [44].

Other imaging modalities such as computed tomography (CT) and magnetic resonance imaging (MRI) are currently limited to ground-based pre- and postflight evaluations. CT imaging is standard of care for evaluating a multitude of medical conditions, like appendicitis and bowel obstruction, as well as a wide range of injuries from trauma. However, radiation doses range from 2 to 20 millisieverts (mSv) depending on protocol and anatomy, and in-flight CT is not feasible due to mass, power, and shielding requirements. MRI has been used extensively to investigate neuro-ocular and cerebrovascular effects pre- and postflight. While the availability of MRI would be ideal and although compact MRI technologies are under development, current systems are not flight-ready due to their large size and mass, electromagnetic shielding requirements, high power consumption, and potential interference with spacecraft avionics. The absence of these modalities reinforces the importance of maximizing diagnostic yield from lower-resource tools and applying clinical reasoning to compensate for incomplete data.

4.4 Medical kits and pharmaceuticals

The medical kits currently flown aboard the ISS are designed for modularity to allow reconfiguration based on mission duration, crew medical profiles, and anticipated clinical scenarios [45, 46]. Items are selected based on long-term stability at ambient storage conditions, low maintenance requirements, and minimal packaging volume [46]. Onboard pharmaceuticals include a variety of analgesics, antiemetics, antimicrobials, anticoagulants, corticosteroids, cardiovascular agents, and a limited selection of sedatives and anesthetics [45]. However, pharmaceutical capabilities are limited due to the inability to manufacture drugs inflight.

Pharmaceutical administration in microgravity presents unique challenges. Gravity-driven intravenous fluid delivery systems cannot be used and infusion must instead be accomplished using pneumatic pressure cuffs applied to flexible fluid reservoirs, which is a technique shown in reduced-gravity experiments to achieve clinically relevant flow rates for both crystalloid and blood-analog fluids [31]. Air-bubble exclusion is critical, and infusion lines must include in-line filters or bubble traps to prevent embolism [31]. More precise infusion rates for medications such as vasopressors or sedatives require small, electrically powered infusion pumps. These devices have been adapted for space and are included in the ISS medical inventory [47, 48]. For exploration-class missions, where the likelihood of encountering medical events increases, these tools must be reliable and intuitive, and treatment decisions must be tailored to limited supplies and the evolving condition of the patient.

Drug degradation in space is an active area of investigation. Pharmaceuticals stored on long-duration missions may be exposed to radiation and temperature fluctuations that could shorten shelf life or reduce efficacy. To mitigate these risks, medication selection emphasizes agents with long shelf lives, low sensitivity to environmental variation, and minimal potential for harmful drug-drug interactions [49, 50]. Repackaging for microgravity use typically includes minimizing headspace in fluid containers and utilizing manufacturer's unit dose packages [51]. During extended missions, maintaining a viable pharmaceutical inventory will require careful tracking, contingency planning, and clinical judgment to determine when substitutions or dose adjustments are appropriate.

5. Ethical considerations

5.1 Ethical dilemmas in resource allocation and clinical decision-making

One of the most sobering questions a space medical officer must face is: What happens when a crew member becomes critically ill or injured, but available resources are insufficient to save them without jeopardizing the survival of others? [52] In deep space, every gram of medical equipment is preplanned, resupply is impossible, and evacuation to Earth is out of the question. On Earth, the principle of beneficence—doing everything possible for the patient—is often paramount. In space, the priority shifts toward a utilitarian ethic: maximizing the survival and operational capability of the crew as a whole [52].

Pre-mission planning is essential. Crews should agree on protocols for emergencies, including the possibility of advanced directives [52]. Some ethicists have suggested explicit, informed consent acknowledging that, in extreme circumstances, a crew member's life might be sacrificed to preserve the mission or the rest of the crew—though NASA has not formally adopted such language. What has been discussed is the prioritization of interventions that offer high benefit-to-cost ratios in terms of supplies, risk, and crew safety.

The prophylactic surgery debate illustrates this tension. To avoid in-flight surgical emergencies such as appendicitis, some have proposed preemptively removing healthy organs (e.g., appendix, gallbladder) before departure [53]. While this could eliminate emergency scenarios, it violates the principle of nonmaleficence ("do no harm") and carries its own surgical risks. However, in certain higher-risk cases, such as the known presence of gallstones, the benefits of prophylactic removal may outweigh the risk of the procedure itself [53].

The principle of justice, meaning fair distribution of resources, takes on new urgency in such contexts. This may parallel battlefield triage, where resources are focused on those with the greatest likelihood of survival [25]. The death of a crew member also presents both technical and ethical challenges [5, 25]. Protocols exist for ISS, including secure containment of remains within hours [52]. .Ethically, dignity and respect for the deceased are paramount, and some form of memorial or ritual may help the surviving crew cope. Resource allocation in space requires balancing autonomy, beneficence, nonmaleficence, and justice in ways that may diverge from terrestrial norms. Clear ethical frameworks established before launch can help ensure these decisions are made with foresight rather than under the emotional strain of crisis.

5.2 Cultural sensitivity, team dynamics, and behavioral health

Ethical medical practice in space extends beyond clinical care to maintaining a cohesive, functional crew. Space missions are inherently multicultural—as seen on the ISS, where American, Russian, European, Japanese, and Canadian astronauts live and work together for months [33]. Each brings unique perspectives on health, privacy, authority, and communication. Recognizing and respecting these differences is critical to crew morale, which is itself a determinant of health and mission success [33].

Cultural differences in medical practice have historically required negotiation. The Multilateral Medical Operations Panel has facilitated integrated medical support plans while allowing some agency-specific flexibility, provided safety is not compromised. Astronauts may need to accept treatments unfamiliar to their own medical culture, and physicians must avoid imposing one agency's practices without evidence.

Communication styles and autonomy also vary across cultures. Some astronauts may be less inclined to question medical authority; others expect shared decision-making [25]. This can affect informed consent, particularly under peer or mission pressure. NASA policy states that no astronaut will be coerced into treatment or research participation, but in practice, refusal can be complicated by crew dynamics. Stigma around mental health remains a concern; ethical responsibility lies in creating an environment where seeking help does not jeopardize career prospects [5].

Behavioral health is a priority on long missions [5]. Phenomena such as the "third quarter effect," when morale dips in the latter half of a mission, have been documented in both spaceflight and analog environments [54]. Depression, irritability, and cognitive fatigue can emerge even in highly trained crews. Countermeasures include regular private conferences with behavioral health professionals, opportunities for recreation, and supportive peer relationships [5, 55].

Team cohesion and conflict resolution are integral to mission safety [25, 33]. Commanders are trained in multicultural leadership and conflict management. Crews establish shared norms—from holiday celebrations to equipment-use schedules—that reinforce inclusion and mutual respect [33]. Language barriers are addressed by ensuring English is used in group contexts and by making efforts to translate informal conversations so no one is excluded [33]. Ultimately, the human factors in space medicine—respect, trust, and psychological support—are as important as technical skills. As one astronaut put it, "*Your crewmates are your family, your patients, and your lifeline.*" Ethical care in space must tend to all three roles simultaneously.

6. The irreplaceable human component

While technology can and should augment space medicine, it cannot replace the adaptability, intuition, and trust embodied in the human physician. As we push beyond low Earth orbit toward the Moon and Mars, the question is not simply whether to bring a doctor, but what kind of physician—equipped with which skills, supported by what systems—can best safeguard the health, performance, and resilience of crews venturing into deep space. Emerging systems such as AI-guided procedural platforms, AR overlays, and real-time physiologic monitoring will be essential companions on exploration-class missions [34, 56, 57]. They can expand the diagnostic reach of a small crew, provide just-in-time refresher training, and support standardized execution of complex tasks. However, the most time-critical, high-stakes elements of care—including emergency stabilization, invasive procedures, and complex decision-making under uncertainty—demand the skill, judgment, and intuition of a trained physician. These are the moments when there is no time to consult a manual, wait for a reply from Earth, or rely solely on automated systems. The difference between survival and catastrophe in such scenarios may hinge on having a human medical expert present.

AI systems like the Intelligent Medical Crew Agent and the Visual Ultrasound Learning, Control, and Analysis Network can guide operators step-by-step through intricate procedures, offering corrective feedback and adjusting to the user's skill level [34]. Such capabilities can significantly enhance crew autonomy, but they cannot account for every variable. In microgravity, anatomy can shift, bleeding patterns are altered, and familiar conditions may present in unfamiliar ways due to fluid redistribution, radiation exposure, and other physiological changes [56]. Experienced physicians adapt in real time, modifying their approach based on a highly complex deductive process that goes beyond algorithmic guidance. This adaptability becomes even more critical when multiple hazards converge, as in a traumatic injury sustained by a crewmember already physiologically compromised by the space environment.

Decision-making in deep space is also inherently contextual and ethical. Supplies are finite, evacuation is impossible, and every primary intervention must be weighed against mission priorities and the needs of the entire crew. As Abramoff and Char argue, an outcome-centered ethic—prioritizing the best achievable results for both patient and mission—may require making choices that differ from terrestrial norms [58]. Determining whether to proceed with a resource-intensive intervention or to withhold treatment unlikely to succeed is best undertaken by a clinician who not only understands the medical facts but also the personalities, resilience, and operational roles of each crewmember.

Trust between the medical officer and the crew is a crucial factor in maintaining both health and mission performance. A physician who trains, works, and lives alongside the crew fosters credibility, encourages early reporting of symptoms, and builds the rapport necessary for compliance with treatment [57]. In the psychologically demanding environment of space—characterized by isolation, confinement, and communication delays—this trust can be an essential countermeasure against stress and conflict. While advanced systems can deliver accurate medical information, they cannot provide genuine empathy or the shared sense of commitment that comes from facing the same risks and hardships.

The optimal approach for exploration-class missions is not to choose between human expertise and advanced medical technologies, but to integrate them in a way that maximizes the strengths of both. AI can enhance the physician's situational awareness, monitor crew health continuously, and present evidence-based recommendations in real time [56, 57]. The physician, in turn, interprets those inputs through the lens of experience, mission context, and human judgment, adapting interventions as conditions change. As Cinelli notes, comprehensive in-flight medical systems that integrate physiological, environmental, and behavioral data become most effective when paired with a clinician capable of interpreting and acting upon that information in real time [56].

Ultimately, the irreplaceable value of the human physician in space lies in the unique combination of procedural skill, rapid adaptive reasoning, ethical judgment, and the capacity to foster trust. These qualities are beyond the reach of current or foreseeable autonomous systems. As missions extend farther from Earth, the presence of a physician who embodies these traits will remain the most reliable safeguard against the unpredictable medical challenges of deep space and the biases of AI.

7. Conclusion

In the decades ahead, integrated medical technologies—ranging from autonomous diagnostic systems to advanced robotics—will dramatically expand the capabilities and independence of space crews, enhancing their ability to respond to both anticipated and unforeseen health challenges. These tools will undoubtedly strengthen contingency planning and mitigate some of the inherent risks of exploration-class missions. Yet, even as systems become more sophisticated, they cannot fully replace the adaptability, clinical judgment, and ethical reasoning of a human physician. The complexity of long-duration spaceflight, with its unpredictable medical scenarios, demands a trained medical professional who can synthesize incomplete information, act decisively in emergencies, and address the human dimensions of care. For missions to the Moon, Mars, and beyond, the fusion of cutting-edge technology with the indispensable presence of a space surgeon offers the most robust path to safeguarding crew health and ensuring mission success.

Author details

Bashar Othman, Sara Rothrock and Peter H.U. Lee*
Brown University, Providence, RI, USA

*Address all correspondence to: peter_lee@brown.edu

References

[347] Patel ZS et al. Red risks for a journey to the red planet: The highest priority human health risks for a mission to Mars. npj Microgravity. 2020;**6**(1):33

[348] Camera A et al. Aging and putative frailty biomarkers are altered by spaceflight. Scientific Reports. 2024;**14**(1):13098

[349] Cermack M. Monitoring and telemedicine support in remote environments and in human space flight. British Journal of Anaesthesia. 2006;**97**(1):107-114

[350] Pool SL, McSwain N. Training space surgeons for missions to the Moon and Mars. In: 52nd International Congress on Aviation and Space Medicine. South Africa: Sun City; 2004

[351] Alfano CA et al. Long-duration space exploration and emotional health: Recommendations for conceptualizing and evaluating risk. Acta Astronautica. 2018;**142**:289-299

[352] Landon LBK, John A, Morissette LG, McTigue KR, Parisi ME, Wu S-C, et al. Assessment of the state of communication delay research in preparation for missions beyond low Earth orbit. In: NASA Technical Memorandum. Moffett Field, California: National Aeronautics and Space Administration, Ames Research Center; 2025

[353] Barratt M, Baker E, Pool S, editors. Principles of Clinical Medicine for Space Flight. 2nd ed. New York: Springer; 2020 936 p. DOI: 10.1007/978-1-4939-9889-0

[354] Lee PHU et al. Factors mediating spaceflight-induced skeletal muscle atrophy. American Journal of Physiology-Cell Physiology. 2022;**322**(3):C567-C580

[355] Whedon GD et al. Mineral and nitrogen metabolic studies on Skylab flights and comparison with effects of earth long-term recumbency. Life Sciences and Space Research. 1976;**14**:119-127

[356] Smith SM et al. Bone metabolism and renal stone risk during International Space Station missions. Bone. 2015;**81**:712-720

[357] Nelson ES, Mulugeta L, Myers JG. Microgravity-induced fluid shift and ophthalmic changes. Life (Basel). 2014;**4**(4):621-665

[358] Heer M, Paloski WH. Space motion sickness: Incidence, etiology, and countermeasures. Autonomic Neuroscience. 2006;**129**(1-2):77-79

[359] Scott JM et al. Leveraging spaceflight to advance cardiovascular research on Earth. Circulation Research. 2022;**130**(6):942-957

[360] Vernice NA et al. Long-term spaceflight and the cardiovascular system. Precision Clinical Medicine. 2020;**3**(4):284-291

[361] Zerwekh JE. Nutrition and renal stone disease in space. Nutrition. 2002;**18**(10):857-863

[362] Yang J-Q et al. The effects of microgravity on the digestive system and the new insights it brings to the life sciences. Life Sciences in Space Research. 2020;**27**:74-82

[363] Marchal S et al. Challenges for the human immune system after leaving Earth. npj Microgravity. 2024;**10**(1):106

[364] Lee AG et al. Spaceflight associated neuro-ocular syndrome (SANS) and the neuro-ophthalmologic effects of microgravity: A review and an update. npj Microgravity. 2020;**6**(1):7

[365] Ong J et al. Spaceflight associated neuro-ocular syndrome (SANS): An update on potential microgravity-based pathophysiology and mitigation development. Eye (London, England). 2023;**37**(12):2409-2415

[366] Lawson BD, Rupert AH, McGrath BJ. The neurovestibular challenges of astronauts and balance patients: Some past countermeasures and two alternative approaches to elicitation, assessment and mitigation. Frontiers in Systems Neuroscience. 2016;**10**:96

[367] Papadopoulos A et al. Space radiation quality factor for Galactic Cosmic Rays and typical space mission scenarios using a microdosimetric approach. Radiation and Environmental Biophysics. 2023;**62**(2):221-234

[368] Desai RI et al. Impact of spaceflight stressors on behavior and cognition: A molecular, neurochemical, and neurobiological perspective. Neuroscience & Biobehavioral Reviews. 2022;**138**:104676

[369] Oluwafemi FA et al. A review of astronaut mental health in manned missions: Potential interventions for cognitive and mental health challenges. Life Sciences in Space Research. 2021;**28**:26-31

[370] Gatti M et al. Affective health and countermeasures in long-duration space exploration. Heliyon. 2022;**8**(5):e09414

[371] Rajput S et al. Medical ethics of long-duration spaceflight. npj Microgravity. 2023;**9**(1):85

[372] Krittanawong C, Singh NK, Scheuring RA, Urquieta E, Bershad EM, Macaulay TR, et al. Human Health during Space Travel: State-of-the-Art Review. Cells. 2022;**12**(1):40. DOI: 10.3390/cells12010040

[373] Russell BK et al. The value of a spaceflight clinical decision support system for earth-independent medical operations. npj Microgravity. 2023;**9**(1):46

[374] Levin DR et al. Enabling human space exploration missions through progressively Earth Independent Medical Operations (EIMO). IEEE Open Journal of Engineering in Medicine and Biology. 2023;**4**:162-167

[375] Doarn CR, Broderick TJ, Pantalos GM, Strangman GE editors. Surgical Capabilities for Exploration and Colonization Space Flight: An Exploratory Symposium. Houston, TX: National Space Biomedical Research Institute; 2016

[376] Iyengar MS et al. Informatics-based medical procedure assistance during space missions. Hippokratia. 2008;**12 Suppl 1**(Suppl 1):23-27

[377] Pantalos GM et al. Intravenous fluid resuscitation capabilities in simulated reduced gravity. Aerospace Medicine and Human Performance. 2023;**94**(8):596-603

[378] Hides J et al. Parallels between astronauts and terrestrial patients—Taking physiotherapy rehabilitation "To infinity and beyond". Musculoskeletal Science & Practice. 2017;**27 Suppl 1**:S32-s37

[379] Duncan JM et al. Organization and management of the International Space Station (ISS) multilateral medical operations. Acta Astronautica. 2008;**63**(7):1137-1147

[380] Buras WR, Hilmers DC. Developing artificially intelligent agents to support earth independent medical capabilities during human exploration-class space missions. npj Microgravity. 2025;**11**(1):51

[381] Cheung HC, De Louche C, Komorowski M. Artificial intelligence applications in space medicine. Aerospace Medicine and Human Performance. 2023;**94**(8):610-622

[382] Mechanic OJ, Persaud Y, Kimball AB. Telehealth systems. In: StatPearls [Internet]. Treasure Island (FL): StatPearls Publishing; 2025

[383] Thirsk R, Williams D, Anvari M. NEEMO 7 undersea mission. Acta Astronautica. 2007;**60**(4):512-517

[384] Doarn CR et al. Evaluation of teleoperated surgical robots in an enclosed undersea environment. Telemedicine and e-Health. 2009;**15**(4):325-335

[385] Pantalone D et al. Robot-assisted surgery in space: Pros and cons. A review from the surgeon's point of view. npj Microgravity. 2021;**7**(1):56

[386] Garcia P et al. Trauma Pod: A semi-automated telerobotic surgical system. International Journal of Medical Robotics and Computer Assisted Surgery. 2009;**5**(2):136-146

[387] Sargsyan AE, Lyndon BJSC. The International Space Station Ultrasound Imaging Capability Overview for Prospective Users, NASA/TP; 2006-213731. Houston, TX: National Aeronautics and Space Administration, Johnson Space Center; 2006

[388] Garcia KM et al. Real-time ultrasound assessment of astronaut spinal anatomy and disorders on the International Space Station. Journal of Ultrasound in Medicine. 2018;**37**(4):987-999

[389] Chiao L et al. Ocular examination for trauma: Clinical ultrasound aboard the International Space Station. American Journal of Ophthalmology. 2005;**140**(4):775-776

[390] Fischetti C et al. Space ultrasound: A proposal for competency-based ultrasound training for in-flight space medicine. The Western Journal of Emergency Medicine. 2024;**25**(2):275-281

[391] Pavy-Le Traon A et al. Pharmacology in space: Pharmacotherapy. Advances in Space Biology and Medicine. 1997;**6**:93

[392] Blue RS et al. Supplying a pharmacy for NASA exploration spaceflight: Challenges and current understanding. npj Microgravity. 2019;**5**(1):14-12

[393] Kopczynski HD. Inflight intravenous therapy: A new dimension. Aviation, Space, and Environmental Medicine. 1975;**46**(10):1305-1308

[394] Horn KJ et al. Insulin delivery pumps for human spaceflight: Steps toward an accessible space future. Wilderness & Environmental Medicine. 2025;**36**(1_suppl):81S-88S

[395] Berman E, Eyal S. Drug interactions in space: A cause for concern? Pharmaceutical Research. 2019;**36**(8):114-116

[396] Daniels VR, Williams ES. Exploring the complexities of drug formulation selection, storage, and shelf-life for exploration spaceflight. British Journal of Clinical Pharmacology. 8 Nov 2023. DOI: 10.1111/bcp.15957

[397] Plante I et al. The long-term stability of solid-state oral pharmaceuticals

exposed to simulated intravehicular space radiation. npj Microgravity. 2025;**11**(1):17

[398] Stepaniak P et al. Operational considerations for crew fatality on the International Space Station. Aerospace Medicine and Human Performance. 2023;**94**:705-714

[399] Ball CG et al. Prophylactic surgery prior to extended-duration space flight: Is the benefit worth the risk? Canadian Journal of Surgery. 2012;**55**(2):125-131

[400] Kanas N, Gushin V, Yusupova A. Whither the third quarter phenomenon? Aerospace Medicine and Human Performance. 2021;**92**:689-691

[401] Jack S. Behavioral Issues Associated with Long-Duration Space Expeditions: Review and Analysis of Astronaut Journals (Experiment 01- E104), Final Report. NASA/TM-2010- 216130. Houston, TX: National Aeronautics and Space Administration, Johnson Space Center; 2010

[402] Cinelli I. The role of artificial intelligence (AI) in space healthcare. Aerospace Medicine and Human Performance. 2020;**91**(6):537-539

[403] Panch T, Mattie H, Celi LA. The "inconvenient truth" about AI in healthcare. npj Digital Medicine. 2019;**2**(1):77

[404] Abramoff MD, Char D. What do we do with physicians when autonomous AI-enabled workflow is better for patient outcomes? American Journal of Bioethics. 2024;**24**(9):93-96

Chapter 11

Perspective Chapter: Technology and Innovation Driven by Space

Kimia Seyedmadani

Abstract

This chapter focused on how space affects technology and innovation. Exploring how the duration and mission design change the internal environmental and also human in the loop design associated with developing medical and countermeasure hardware in spaceflight. In addition to the hardware what internal infrastructure such as interfaces suggested and possible computing capabilities. As the AI/ML technology increasing and cloud based system how to approach long distance and surface missions while access to satellite are limited and almost not possible. At the end we will look at how some of these design may inform global health technology and improve overall life on earth.

Keywords: mission, AI, biomedical, countermeasure, innovation

1. Introduction

The "Eureka" moment that drives innovation does not follow a fixed formula; however, space exploration has repeatedly demonstrated its role as a fertile environment for innovative ideas and developments, largely driven by the necessity and uniqueness of the challenges encountered. Challenges in space exploration are governed by space stressors and design reference missions. Design Reference Missions (DRMs) provide a structured framework that defines mission scope, duration, and operational constraints essential for planning human space exploration. A central element within DRMs is *mission duration*. Mission duration influences not only environmental factors but also every aspect of hardware, including redundancy, autonomy, and reliability specifications, thereby impacting systems ranging from life support to data management.

Long-duration missions, such as Mars expeditions, exemplify these challenges. NASA's *OCHMO-TB-007: Mission Duration* [1] highlights key constraints including:

- Substantial *communication delays* (~40 minutes round-trip latency),
- Complete *absence of evacuation and resupply options*, requiring highly reliable and autonomous onboard resources, and
- A crucial need for *autonomous medical diagnostics and treatment* given Earth-based intervention delays [1].

These factors necessitate fault-tolerant system designs, closed-loop recycling of consumables like water and air, and advanced AI-enabled fault detection and correction. Furthermore, mission duration impacts pharmaceutical stability, nutrition packaging, and maintenance schedules, cascading across mission system design [1].

Overlaying these temporal design considerations are the unique environmental challenges of spaceflight that further influence innovation trajectories. Chief among these are:

- *Microgravity*, which induces profound physiological effects such as muscle atrophy, bone density loss, and sensorimotor disturbances. Addressing these requires advanced exercise equipment, pharmaceutical countermeasures, and life support systems capable of sustaining astronaut health on prolonged missions [2, 3].

- Exposure to *space radiation* from galactic cosmic rays and solar particle events, which threatens crew health and mission electronics alike, demanding development of enhanced shielding materials, radiation-hardened electronics, and biological countermeasures [4, 5].

- The *vacuum of space* imposes challenges including material outgassing and thermal management under extreme temperature swings, requiring innovative materials and thermal control solutions [6].

- *Communication latency*, especially pronounced in deep-space contexts, compels reliance on autonomous systems and onboard AI to maintain safe and efficient operations without continuous Earth support [1, 7].

Strategic research demonstrates that mission sequencing and launch frequency strongly affect the innovation pipeline. High upfront costs and infrequent missions deepen the "Valley of Death" for technology development, making smaller, more frequent missions advantageous for accelerating innovation maturation and return on investment [7].

Biomedical research led by the National Space Biomedical Research Institute (NSBRI) complements these engineering efforts by focusing on mitigating physiological risks posed by long-duration spaceflight—addressing musculoskeletal, cardiovascular, neurological, radiation, and psychological challenges through integrated countermeasures [8]. NSBRI's education and outreach programs also cultivate a skilled workforce and public support, critical for sustaining long-term exploration goals [9].

NASA's Biological and Physical Sciences Decadal Surveys further highlight these themes, prioritizing research areas such as radiation risk reduction, muscle and bone health, autonomous medical care, and closed-loop life support—fundamental for the success of extended human missions beyond Earth orbit [10].

Together, these interconnected challenges defined by mission duration and space environment conditions create a complex innovation landscape. Progress in engineering, biomedical science, and human-centered design is imperative—not only for mission success but also for advancing technologies with transformative applications on Earth **Figure 1**.

Figure 1.
Apollo astronaut collecting sample on the surface of the moon with every day geology tools that was modified to be used with gloves of crew, bringing innovation to existing technology per new needs. (Picture credited by NASA).

2. Impact of DRM on medical and countermeasure hardware

Design Reference Missions (DRMs) define mission parameters such as duration, destination, crew size, and risk tolerance, which directly shape the development, certification, and operation of medical and countermeasure hardware. To ensure astronaut health and mission success in extended spaceflight, integrated environmental control systems, autonomous medical technologies, human-in-the-loop design, adaptive certification, and data-driven risk management are critical.

2.1 Environmental control and hardware specifications

Extended-duration missions, including lunar surface stays and Mars transit, require closed-loop life support and durable medical devices that can operate autonomously under mass, volume, and power constraints. These systems must endure radiation exposure and variable gravity conditions. Medical hardware such as biosensors, diagnostic tools, and medication delivery systems must provide intuitive guided workflows for use under stress and limited dexterity by non-expert crew [11–13].

The Human System Drivers for Exploration Missions NASA report notes medical resource requirements as significant mass-volume drivers; for example, a 386-day near-Earth object (NEO) mission requires an estimated 62 kg and 0.15 m^3 of medical kit mass and volume [14]. Such planning accounts for probabilistic crew evacuation and loss risks, emphasizing adequate hardware sizing and redundancy.

2.2 Human in the loop system design and human-system risk management

Human-in-the-loop design integrates astronaut feedback to mitigate human-system risk. NASA employs directed acyclic graphs (DAGs) to model causal links from environmental hazards to health outcomes, guiding the development of interactive systems like the ISS Human Research Facility-1 (HRF-1). This approach addresses failure modes such as operator error and cognitive overload with adaptive displays, real-time biometrics, context-aware alerts, and embedded clinical decision support [14, 15].

The Human System Risk Board (HSRB) continually updates risk models based on analog missions, parabolic flights, and in-flight data. These efforts address mission-limiting factors including radiation effects, intracranial pressure, bone demineralization, behavioral health, and decompression sickness, all critical in DRM planning [13].

2.3 Adaptive certification approaches

Certification frameworks for medical hardware adapt terrestrial standards to spaceflight-specific conditions: increased radiation, communication latency, autonomy requirements, and tolerance for partial system degradation. Phased validation spans ground analogs (NEEMO, HERA), ISS demonstrations, and lunar testbeds. Modular certification paths support emerging technologies such as AI diagnostics and wearable monitors, with iterative risk-based evidence gathering to ensure safety [13, 16]. Hardware certification also considers logistical and operational drivers, including consumable shelf life, onboard medical expertise, abort scenarios, and EVA support systems such as suit-port versus airlock designs [14].

2.4 Countermeasure systems: Exercise, medical kits, and health monitoring

Countermeasure systems are vital to mitigate the physiological effects of microgravity and mission stresses over extended durations. NASA's exercise countermeasures evolved from the Apollo-era rope friction devices through Skylab's Mini-Gym and MK-2 to Space Shuttle cycle ergometers and treadmills. On the ISS, advanced systems such as the Advanced Resistive Exercise Device (ARED), Combined Operational Load-Bearing External Resistance Treadmill (COLBERT), and Cycle Ergometer with Vibration Isolation and Stabilization System (CEVIS) maintain musculoskeletal and cardiovascular health, supported by vibration isolation technology to protect spacecraft integrity [17, 18].

Medical kits aboard the ISS—housed within the Crew Healthcare Systems (CHeCS)—include Advanced Life Support Packs (ALSP), Ambulatory Medical Packs (AMP), Crew Contamination Protection Kits (CCPK), Respiratory Support Packs (RSP), and Ancillary Support Packs (HASP). These kits utilize RFID tagging for inventory control and rapid access. Portable Clinical Blood Analyzers (PCBA) monitor critical biomarkers such as urine calcium to track bone health, with ongoing efforts to validate these measures in microgravity [19].

Interdisciplinary collaboration among National Aeronautics and Space Administration (NASA) divisions and external institutions like the University of Southern California and the Food and Drug Administration (FDA) fosters innovation in health maintenance systems, ensuring terrestrial medical advances translate to spaceflight hardware and procedures [19].

2.5 Strategic countermeasure planning for Artemis missions

As NASA embarks on Artemis missions to the Moon and eventually Mars, a strategic, risk-informed approach to countermeasures is essential. Key considerations include radiation mitigation, microgravity-induced physiological degradation, psychological health support, and medical system autonomy to reduce reliance on Earth-based intervention [20]. Proactive planning leverages lessons learned from Apollo, Sky lab, ISS and Shuttle programs, integrating advanced exercise regimes, sophisticated health monitoring, and autonomous medical support systems. These initiatives aim to sustain astronaut health, performance, and safety amid the increased duration and isolation of deep space missions [20].

3. Internal infrastructure and interface innovations

The advancement of internal infrastructure and interface technologies within space exploration platforms is pivotal to supporting comprehensive innovation in biomedical capabilities in extraterrestrial environments. As mission durations extend and operational complexity increases, the need for autonomous, reliable, and sophisticated biomedical systems becomes critical to ensuring crew health and mission success [21].

Internal Infrastructure Innovations focus on the integration of life-support systems, medical devices, and computational resources that can operate under the constraints of limited volume, mass, and energy availability inherent to spacecraft and planetary habitats. These systems must provide continuous monitoring and support, including vital signs assessment, diagnostic capabilities, and therapeutic interventions, often with minimal direct input from Earth-based medical teams due to communication delays [22].

Interface Innovations emphasize human-centered design principles to optimize usability and cognitive ergonomics in high-stress, confined environments. Advanced human-machine interfaces (HMI) facilitate seamless interaction between crew members and onboard medical systems, enabling efficient data visualization, decision support, and remote operation capabilities. Interfaces increasingly incorporate multimodal feedback, including touch, voice, and augmented reality (AR), to support complex medical procedures under microgravity or partial gravity conditions [23].

A notable domain within biomedical capabilities is the enhancement of *surgical interventions* in space. Microgravity presents unique challenges to traditional surgical techniques, such as fluid behavior and tissue handling. To address these, the development of miniaturized, robotic-assisted surgical systems tailored for space environments is underway. These systems leverage precision control and advanced imaging to facilitate minimally invasive procedures, reducing physiological trauma and recovery time [24]. Moreover, tele-operated surgical platforms combined with AI-driven guidance algorithms can compensate for limited onboard surgical expertise and communication delays, enabling remote surgical support and decision-making [25].

Artificial Intelligence (AI) integration is a transformative element across the internal infrastructure spectrum. AI algorithms enhance diagnostic accuracy, predict medical events, personalize countermeasure regimens, and optimize resource management. Machine learning models trained on vast terrestrial and spaceflight medical datasets provide predictive analytics that can identify early signs of conditions such as spaceflight-associated neuro-ocular syndrome (SANS) or musculoskeletal deconditioning [26]. Furthermore, AI-powered virtual assistants support autonomous medical operations, guiding crew through complex procedures, interpreting sensor data, and managing emergency responses without real-time ground support [27].

In summary, the synergistic development of robust internal infrastructure and intuitive interfaces, augmented by AI and robotic surgical technologies, constitutes a critical frontier in space exploration biomedical capabilities. These innovations not only enable effective medical care in extreme environments but also offer promising translational applications for remote and resource-limited healthcare on Earth.

4. Environmental and human-centered habitat design

Designing closed habitats—such as spacecraft, rovers, and planetary bases—demands a comprehensive, integrated approach that merges physiological life support, human-centered design, environmental adaptation, and microbial management. These systems must not only sustain life but actively support crew well-being, performance, and psychological resilience under isolation and confinement.

4.1 Design thinking and user participation

Human-centered design thinking—featuring empathy mapping, stakeholder engagement, and iterative prototyping—is central to modern habitat development. Programs such as HERA, CHAPEA, and Lunar Gateway use VR/AR mockups and analog feedback loops to refine spatial layout, lighting schemes, acoustic zones, and privacy features. Involving end-users early and continuously improves both functionality and crew satisfaction [28, 29].

4.2 Adaptive and rhythmic architecture

The *Rhythmic Buildings Strategy* integrates environmental and operational rhythms—circadian cycles, thermal variations, and crew schedules—into habitat design. Bridgens et al. applied a 33-parameter framework to conceptual Martian habitats in Jezero Crater, enabling adaptive lighting modulation and thermal zonation [30]. Smitherman & Griffin's AIAA concept advocates modular, radiation-tolerant deep-space habitats optimized for usability and protection [31]. Similarly, foundational research highlights the complex environmental conditions on Mars and the Moon—radiation, atmospheric differences, temperature extremes—that require innovative architectural solutions [32].

4.3 Artificial gravity considerations

Artificial gravity systems—achieved through rotating habitats or centrifugation—offer a promising approach to counteract the physiological challenges of microgravity, including muscle atrophy, bone demineralization, vestibular deconditioning, and cardiovascular deconditioning. Integrating artificial gravity significantly influences habitat architecture and mechanical systems, necessitating careful design to balance rotational forces with crew usability and comfort.

- *Physiological Foundations*: Young [33] provides an in-depth review of microgravity-induced physiological deficits and highlights how sustained artificial gravity can simulate terrestrial gravity to preserve musculoskeletal, vestibular, and cardiovascular health [34].

- *Structural and Design Innovations*: Majji et al. [35] propose a modular, growth-capable rotating habitat based on tensegrity structures. Their design minimizes mass while maintaining centrifugal forces appropriate for human habitation, incorporating active control for spin stability and planning for agricultural integration in low-gravity zones [35]. Chen et al. further demonstrate the feasibility of scalable tensegrity-based rotating habitats, detailing how such systems can be expanded during prolonged missions while maintaining a stable artificial-gravity environment [36–38].

- *Psychological and Perceptual Benefits*: Exposure to artificial gravity can also enhance psychological well-being by offering familiar gravitational cues, improving spatial orientation, and reducing motion-related discomfort and disorientation. Human factors studies on rotating environments underscore the importance of minimizing gravity gradients and rotation-induced vestibular stressors [39].

4.4 Integrated environmental systems

- *Air Quality and Lighting:* Adaptive solid-state LED systems, calibrated for intensity and spectral composition, promote circadian alignment, improving sleep, alertness, and mood during ISS missions and analog trials [40, 41]. A dynamic lighting schedule mimicking natural light patterns was shown to significantly enhance neurobehavioral performance during a 45-day simulated space mission, underscoring circadian-informed lighting's critical role [42].

- *Nutrition and Food-Microbiome Interactions:* Diets tailored for long-duration missions must support immune, musculoskeletal, and neurobehavioral health. Fresh produce systems like ISS *Veggie* offer nutritional and psychological benefits [43, 44]. Research indicates that food composition drives microbiome diversity and metabolite production—fiber-rich, complex diets foster beneficial short-chain fatty acid production, while highly processed, low-fiber diets can lead to dysbiosis and inflammation. Precision food design, accounting for macronutrient ratios and food structure, is thus essential to support microbial and host health in closed environments [45–47].

- *Biophilia and Sensory Enrichment:* Incorporating plants, natural textures, and fractal patterns reduces stress, restores cognition, and mitigates isolation. Interactions with greenery in analog habitats have shown notable improvements in crew mental state [48, 49].

- *VR Enrichment:* Immersive VR experiences—ranging from nature to art—alleviate sensory monotony and asthenia in isolation, enhancing mood and cognitive function [50, 51].

5. Microbiome monitoring and control

Closed habitat microbiomes evolve over time. A 105-day Lunar Palace simulation noted shifts in salivary microbiota and inflammatory cytokines, suggesting immune modulation. Similarly, MARS500 and HI-SEAS missions recorded microbial convergence and potential dysbiosis, underscoring the need for continuous microbial monitoring and tailored interventions [52–56].

5.1 Ergonomics and spatial flexibility

Habitats now incorporate modular, transformable interiors—adjustable partitions, acoustic zones, multifunctional living modules—guided by analog mission feedback. Computational tools optimize these layouts for mass, volume, and task efficiency, improving autonomy, privacy, and crew dynamics [45, 57].

5.2 Human-centered interfaces

Effective interface design in space habitats plays a critical role in safety, performance, and usability—especially in high-stakes or microgravity environments.

- *Ergonomics and Accessibility:* Interfaces, control panels, and wearables are optimized for zero or partial gravity environments. Designs account for suited operation (e.g., gloved hands), body orientation, and reduced motor control. Tactile feedback, accessible reach zones, and modular placement improve interaction under constrained movement [58, 59].

- *Cognitive Load and Situational Awareness:* Interfaces are increasingly adaptive and context-aware, capable of prioritizing essential data and supporting decision-making in time-critical operations. Smart displays and multimodal feedback (visual, auditory, haptic) help reduce mental workload and increase mission efficiency [60, 61]. Studies have shown that intuitive dashboards and assistive AI systems can reduce error rates and improve crew autonomy in analog simulations [62].

6. Cognitive support and psychological well-being

In addition to environmental controls, habitat design increasingly focuses on supporting cognitive function and psychological resilience. Zhang and Surampalli emphasize adaptive environmental systems—such as lighting and acoustics—that dynamically respond to crew activity and circadian needs, reducing fatigue and mental strain [63]. Complementing these, Johnson et al. explore how immersive and restorative environments, combined with interface design, enhance mood and reduce stress in analog habitats [64]. Together, these studies reinforce the importance of integrated cognitive support systems to maintain crew performance during prolonged missions.

7. Harnessing artificial intelligence for medical diagnosis and treatment during space exploration missions

Artificial Intelligence (AI) plays an increasingly vital role in ensuring the health and safety of astronauts on long-duration space missions. In environments where immediate expert intervention is not feasible, AI-driven systems provide autonomous or semi-autonomous support for diagnosis, treatment, and medical decision-making

These systems not only help mitigate communication delays but also compensate for the limited medical expertise onboard.

7.1 AI-enabled clinical decision support

NASA's Exploration Medical Capability (ExMC) initiative has advanced the deployment of Clinical Decision Support Systems (CDSS) tailored to spaceflight conditions. These systems integrate real-time physiological data, diagnostic imaging, and environmental inputs to guide treatment pathways [65]. CDSS tools enable onboard crew—often non-expert medics—to make informed decisions using context-aware recommendations, reducing dependence on Earth-based medical teams.

7.2 Autonomous medical officer support (AMOS)

The Autonomous Medical Officer Support (AMOS) platform represents a leap forward in autonomous healthcare delivery. It equips crew members with the ability to conduct self-guided diagnostic procedures, including ultrasound imaging, while receiving real-time feedback from AI-based coaching systems [66]. This reduces the cognitive and technical burden on astronauts during medical emergencies and facilitates just-in-time training.

7.3 Communication-aware HITL AI systems

A major challenge in deep space is communication latency, which can span minutes to hours depending on mission distance. To address this, Mavrakis et al. [67] propose Human-in-the-Loop (HITL) AI communication frameworks that enable asynchronous yet contextually aware interactions between astronauts and remote experts. These systems enhance the quality and clarity of medical consultations despite signal delays, using intelligent agents to summarize, rephrase, and flag critical information for remote review.

7.4 Advanced medical imaging enhancement via AI

Recent advancements in medical image processing, such as the retinex-based image enhancement combined with lightweight deep learning models, have demonstrated significant improvements in diagnostic accuracy [68]. While originally developed for dental periapical lesion detection, these techniques have broad applicability for space medicine, where high-quality imaging is critical yet challenging due to limited hardware capabilities and environmental factors. Integrating such AI-powered image enhancement can improve clarity in ultrasound or other onboard imaging modalities, thereby facilitating more accurate diagnostics in resource-constrained settings.

7.5 AI centaurs: Hybrid human-AI diagnostic systems

Babocs and Stark [69] introduce the concept of AI centaurs—hybrid systems that synergize AI computational strengths with human expertise for differential diagnosis in spaceflight medicine. This approach leverages AI's ability to analyze large datasets and detect subtle patterns, while maintaining the astronaut's or medic's contextual knowledge and judgment in the decision loop. The AI centaur model enhances diagnostic accuracy and response times, which is crucial for managing complex or ambiguous medical conditions autonomously during missions.

7.6 Trustworthy and ethical AI integration

For AI systems to be viable in mission-critical healthcare scenarios, they must operate within strict ethical boundaries. Trustworthy AI principles—such as transparency, reliability, and human oversight—are central to ensuring safety and accountability [70]. HITL paradigms preserve human authority over final medical decisions, ensuring that AI recommendations remain assistive rather than directive.

7.7 AI limitations onboard space missions

Despite significant advancements, AI systems deployed during space missions face inherent limitations. Computational constraints aboard spacecraft restrict the complexity and size of AI models, necessitating lightweight architectures that may compromise diagnostic accuracy or decision depth [68]. Furthermore, the scarcity of high-quality, mission-specific training data poses challenges for generalizing AI performance across diverse medical scenarios in space. Environmental factors, such as radiation exposure and hardware degradation, may impact system reliability and increase failure risks. Crucially, AI systems cannot fully replace human intuition and contextual judgment, underscoring the importance of Human-in-the-Loop designs to mitigate risks arising from AI errors or ambiguities [69, 70].

7.8 Future directions: Agentic AI and edge autonomy

The convergence of large language models (LLMs), agentic AI, and edge-compute infrastructure is shaping the future of autonomous space medicine. NASA's roadmap includes the integration of mission-tuned AI agents capable of managing diagnostics, querying databases, orchestrating treatment workflows, and interacting naturally with crew members [71]. These agents are designed to function offline, ensuring robustness even in the event of communication blackouts. The application of AI in space medicine holds transformative potential for diagnosing and treating conditions in remote, resource-limited environments. Systems like CDSS and AMOS, coupled with intelligent communication agents, advanced imaging enhancement, AI centaurs, and ethical HITL frameworks, offer a pathway to sustained astronaut health beyond Earth orbit. Continued interdisciplinary research and validation—such as that demonstrated in *SpaceCHI 2025* and recent AI imaging and centaur models—will be essential to transition these technologies from prototypes to mission-critical assets.

8. AI, machine learning, and cloud systems driven by planetary mission demands

8.1 AI/ML innovations spurred by intermittent connectivity

Planetary missions impose stringent constraints on communication due to vast distances, resulting in intermittent connectivity, high latency, and limited bandwidth. These limitations have driven significant innovation in AI and machine learning to enable onboard autonomy. Traditionally, data processing and decision-making relied heavily on uplink/downlink communication with Earth-based ground stations. However, the need for real-time responsiveness during critical mission phases—such as landing, surface navigation, or anomaly response—has accelerated the development of advanced AI hardware capable of onboard inference. Devices like the Movidius Myriad X vision processing unit exemplify this trend, providing efficient, low-power AI computation directly on spacecraft [72–75]. Moreover, the complexity of planetary environments has encouraged the adaptation of reinforcement learning architectures, including GPT-scale models like SigmaZero on the CAPSTONE™ platform, to support autonomous navigation and decision-making under uncertain and dynamic conditions in cislunar and deep space contexts [76, 77].

8.2 Autonomy and decision support: AI innovation for mission resilience

The inability to maintain continuous communication during planetary missions has necessitated AI systems capable of autonomous fault detection, anomaly management, and health monitoring. This demand has accelerated the creation of AI algorithms that can perform sophisticated diagnostics and self-healing without ground intervention. JAXA's Epsilon rocket serves as a pioneering example, leveraging AI to conduct pre-launch vehicle health assessments that reduce preparation time and increase operational agility [78]. Similarly, the integration of deep anomaly detection and adaptive self-repair mechanisms into mission architectures reflects the imperative for AI systems to maintain spacecraft health and mission continuity in environments where human intervention is delayed or impossible [79–81]. These innovations address mission-critical needs, ensuring reliability during long-duration planetary exploration.

8.3 Distributed and heterogeneous computing architectures

The challenges of planetary exploration—such as limited power availability, harsh environments, and communication constraints—have driven the development of distributed computing architectures that strategically balance onboard and ground-based processing. Hybrid architectures are evolving to distribute workloads between spacecraft, surface assets, orbiters, and Earth-based cloud infrastructure. Edge AI accelerators embedded in landers and rovers facilitate immediate data processing, enabling rapid situational awareness and decision-making. Meanwhile, intermittent data offloading to orbiters or cloud systems allows for deeper analysis and mission replanning during communication windows. To meet these demands, recent AI innovation emphasizes heterogeneous co-processing architectures featuring multiprocessor systems-on-chip (MPSoCs), vision and tensor processing units (VPUs/TPUs), and field-programmable gate arrays (FPGAs). These designs enable optimized trade-offs between computational power, energy efficiency, and mission-specific adaptability, reflecting the unique operational requirements of planetary missions [72, 75, 82, 83].

8.4 Driving AI innovation: The role of advanced models in planetary missions

The demanding environment of planetary exploration has catalyzed the integration of advanced AI models, including large-scale neural networks such as GPT, into mission systems. These models represent a leap forward in enabling onboard autonomy, complex data interpretation, and human-machine collaboration. AI assistants inspired by such models can autonomously synthesize scientific data, generate mission-critical reports, and propose adaptive strategies that respond to evolving mission scenarios without waiting for Earth-based commands. This capability reduces operational delays and empowers spacecraft to undertake more complex tasks independently. The unique constraints of planetary missions—communication delays, limited computational resources, and dynamic environments—have thus been a major driving force behind AI innovation, pushing the development of robust, efficient, and adaptable AI systems that are transforming the future of space exploration.

9. Integrated research and development

The integrated approach to research and development in human and biomedical domains is fundamental for ensuring astronaut health, safety, and operational effectiveness during long-duration space missions. NASA's Human Research Roadmap serves as a comprehensive strategic framework that identifies critical knowledge gaps related to spaceflight-induced stressors, including the combined physiological and psychological effects of microgravity, radiation, and isolation. To translate these priorities into actionable research, NASA's Research Operations and Integration Group (ROIG) plays a pivotal role by coordinating multidisciplinary research activities across various centers and flight programs, overseeing resource allocation, scheduling, and risk management, thereby ensuring alignment with mission objectives and enhancing collaboration between NASA, commercial partners, and international stakeholders. Complementing this, the International Space Station Medical Project (ISSMP) manages the Human Research Facility (HRF) aboard the ISS, providing standardized infrastructure, specialized equipment, and operational support necessary for conducting high-quality biomedical investigations in microgravity. ISSMP ensures astronaut health monitoring, real-time communication between flight crews and ground-based medical teams, and adherence to safety and scientific standards, thus enabling adaptive research protocols and accelerating the development of effective countermeasures and personalized medical interventions to safeguard astronaut health [84].

Similarly, the European Space Agency (ESA) coordinates human research through its Biomedical Office and European Astronaut Centre (EAC), which facilitate the planning, implementation, and integration of biomedical experiments. The EAC provides operational support, astronaut health monitoring, and acts as a liaison for European astronaut training and research activities aboard the ISS and other missions. ESA's multidisciplinary approach fosters collaboration across member states and international partners to ensure robust human research integration [85].

The Japan Aerospace Exploration Agency (JAXA) operates the Human Spaceflight Technology Directorate and manages the JAXA Human Research Center, which oversee Japanese astronaut health research and biomedical experimentation on the ISS. JAXA emphasizes real-time data integration, health monitoring, and countermeasure development, ensuring that research aligns with mission demands and contributes to the global body of human spaceflight knowledge [86].

Biomarker research in spaceflight encompasses a wide array of physiological and molecular indicators. Key biomarkers include cortisol, which reflects stress-related hypothalamic-pituitary-adrenal (HPA) axis activity; inflammatory cytokines such as interleukin-6 (IL-6) and tumor necrosis factor-alpha (TNF-α), indicative of immune system activation; and oxidative stress markers like reactive oxygen species and glutathione levels [87]. These biomarkers have been studied extensively in missions such as the International Space Station (ISS) expeditions, the NASA Twins Study, and Space Shuttle missions, providing valuable insight into human adaptation to microgravity and radiation exposure [88, 89]. Neurotransmitter metabolites have been analyzed to understand cognitive and mood alterations under microgravity conditions, while telomere length and telomerase activity, notably studied during the NASA Twins Study, offer insight into cellular aging processes experienced during long-duration missions [89]. Emerging areas of research focus on microRNAs and epigenetic modifications, which offer promising avenues for personalized medical monitoring and intervention strategies. Together, these biomarkers provide a critical foundation for the development of tailored countermeasures that optimize astronaut health and performance in extreme environments.

In parallel, habitat resilience modeling has emerged as a vital tool for sustaining life support and operational systems in extraterrestrial environments. Simulation platforms such as HabSim enable detailed modeling of habitat dynamics under various disruptive scenarios, including equipment failures, environmental hazards, and unexpected system interactions. These models not only focus on technical robustness but also incorporate adaptive systems capable of responding autonomously to emergent issues, thereby supporting continuous operation with minimal human intervention. Crucially, habitat modeling extends beyond isolated system performance to encompass the complex ecosystem interactions within the habitat, including bioregenerative life support components such as plants, microbial communities, and waste recycling processes. This holistic ecosystem approach facilitates the study of feedback loops, nutrient cycling, and environmental control, allowing researchers to predict how biological and mechanical systems co-adapt over time under the unique stresses of space environments. The integration of resilience and adaptation principles in habitat models enables the development of self-sustaining habitats that can dynamically respond to internal and external perturbations, ensuring mission success during long-duration deep space exploration [90].

10. Translational impact: Space-inspired innovations for global health

10.1 Spin-offs from space hardware and software

The legacy of space exploration programs, most notably the Apollo program, has catalyzed a broad spectrum of technological innovations with profound impacts on global health and everyday life. More than 1800 documented spin-offs have emerged from aerospace research and development, including cordless power tools, fire-resistant suits, advanced heart monitoring systems, and photovoltaic solar panels [91–94]. Noteworthy examples include memory foam, originally developed for astronaut cushioning and now widely used in medical and consumer products; cordless power tools derived from lunar soil collection drills, which led to consumer devices such as the DustBuster vacuum; advanced water purification systems adapted for clean drinking water in remote and disaster-affected areas; and fuel cell technologies that supply sustainable power on Earth [95–99]. Additionally, wireless medical devices benefiting from NASA's spacesuit and battery innovations have enabled minimally invasive surgical procedures with improved patient outcomes. These innovations largely stem from aerospace-grade integrated circuits and highly reliable software systems initially designed for the stringent demands of space missions. Such technologies have found transformative applications in medical devices, emergency response equipment, and sustainable energy solutions on Earth, underscoring the far-reaching benefits of investments in space technology.

10.2 Remote healthcare and diagnostics

Telemedicine technologies pioneered for space missions, where medical expertise is often inaccessible, have been adapted to enhance healthcare delivery in remote and underserved regions worldwide. Miniaturized biosensors, autonomous health monitoring systems, and low-latency diagnostic platforms originally engineered for space telemedicine have been deployed in rural clinics, maritime vessels, and disaster zones to provide timely and accurate point-of-care diagnostics [100, 101]. These systems facilitate continuous patient monitoring and rapid clinical decision-making without the need for immediate specialist intervention, thereby reducing healthcare disparities and improving outcomes in challenging environments. The ability to remotely assess patient health metrics and detect anomalies early is critical for managing chronic conditions and emergency cases in areas with limited medical infrastructure. Moreover, the compact design and energy efficiency required for space hardware ensure these devices are practical and durable for field deployment.

10.3 Additive manufacturing and in situ production

Advances in additive manufacturing protocols developed for space applications, particularly in-situ manufacturing aboard low Earth orbit (LEO) platforms and beyond, have opened promising avenues for revolutionizing medical supply chains on Earth. Techniques refined for fabricating critical components and tools in microgravity environments have enabled the development of novel materials and extreme alloys that are difficult or impossible to produce under terrestrial conditions [102, 103]. Microgravity allows for reduced defects, more uniform microstructures, and unique crystalline growth, which contribute to superior mechanical and biocompatible properties essential for medical implants and prosthetics. For example, NASA's experiments aboard the International Space Station (ISS) have demonstrated the successful 3D printing of advanced titanium alloys and biocompatible polymers with enhanced strength and fatigue resistance [104, 105]. These materials are ideal for producing custom implants and modular medical kits that can be rapidly fabricated on demand during emergencies or in remote locations. This capability is particularly vital for disaster relief and battlefield medicine, where conventional logistics are often hindered by delays and supply chain disruptions. The transfer of space-based 3D printing technologies and material innovations to terrestrial healthcare systems promises enhanced responsiveness, reduced costs, and increased self-sufficiency in medical device availability. Furthermore, additive manufacturing in space supports sustainability by reducing waste and enabling repair or replacement of worn parts without the need for large inventories.

10.4 Broader impact of space research on health technologies

Beyond hardware manufacturing and remote diagnostics, space research has fostered the development of complementary technologies with direct applications in global health. Virtual reality (VR) platforms, initially explored for astronaut mental health and training, have found roles in therapeutic interventions and pain management on Earth. VR therapy assists patients suffering from post-traumatic stress disorder (PTSD), anxiety, and chronic pain by creating immersive environments for exposure therapy and distraction techniques [106]. Closed-loop life support systems developed for maintaining astronaut health in sealed environments inform innovations in sustainable environmental control and patient care facilities, improving air quality and hygiene in hospitals [107]. Furthermore, portable Internet of Things (IoT)-enabled diagnostic devices, designed for continuous monitoring of astronaut health parameters, are now instrumental in remote medicine and disaster response scenarios. These devices facilitate real-time health data transmission and predictive analytics to anticipate medical emergencies [108]. Additionally, design processes developed to create innovative biomedical hardware for spaceflight are proving invaluable for Earth applications, enabling the rapid prototyping and deployment of robust, compact, and multifunctional medical devices suitable for extreme and resource-limited environments [109]. Collectively, these space-derived technologies contribute to enhancing healthcare accessibility, quality, and resilience in diverse terrestrial settings. Recent comprehensive reviews underscore the critical role of space technology innovation in advancing global health outcomes, demonstrating substantial returns on investment through improved healthcare delivery and technological spinoffs [103].

11. Conclusion

This worked focused on showing some of technologies and innovation that were developed under space flight stressors as the result of human in exploration. As we go further away from the planet earth the complexity of the task and harshness of environment drives innovation beyond our imagination, as shown in several section about new technologies and spin off impacted human life quality on earth today. One only can dream of how innovation from Mars exploration will impact the future of mankind.

Acknowledgements

Author is NASA employee, but the views and opinions expressed herein are not necessarily those of NASA or the United States Government.

Conflict of interest

The authors declare no conflict of interest.

Author details

Kimia Seyedmadani
Johnson Space Center, Houston, Texas, USA

*Address all correspondence to: kimia.seyedmadani@nasa.gov

References

[405] NASA. OCHMO-TB-007: Mission duration. In: NASA Technical Brief. USA; 2023. [Online]. Available from: https://www.nasa.gov/wp-content/uploads/2023/12/ochmo-tb-007-missionduration.pdf

[406] McPhee JC, White RJ. Physiology, medicine, long-duration space flight and the NSBRI. Acta Astronautica. 2003;**53**(4-10):239-248. DOI: 10.1016/s0094-5765(03)80001-2

[407] NASA. Microgravity Research. USA: NASA Biological and Physical Sciences Decadal Survey; 2017. [Online]. Available from: https://science.nasa.gov/biological-physical/resources/decadal-surveys/

[408] NASA. Radiation Risks for Space Exploration. USA: NASA Biological and Physical Sciences Decadal Survey; 2017. [Online]. Available from: https://science.nasa.gov/biological-physical/resources/decadal-surveys/

[409] Durante M, Cucinotta FA. Physical basis of radiation protection in space travel. Reviews of Modern Physics. 2011;**83**(4):1245-1281

[410] Mermer E, Ünal R. Passive thermal control systems in spacecrafts. Journal of the Brazilian Society of Mechanical Sciences and Engineering. 2023;**45**(3):160

[411] Szajnfarber Z. Space science innovation: How mission sequencing interacts with technology policy. Space Policy. 2014;**30**(2):83-90

[412] McPhee JC, White RJ, et al. The National Space Biomedical Research Institute's education and public outreach program: Working toward a global 21st century space exploration society. Acta Astronautica. 2009;**65**(7-8):1180-1189. DOI: 10.1016/j.actaastro.2009.02.016

[413] NASA. NASA Biological and Physical Sciences Decadal Survey: Muscle and Bone Health. USA; 2017. [Online]. Available from: https://science.nasa.gov/biological-physical/resources/decadal-surveys/

[414] NASA. NASA Biological and Physical Sciences Decadal Survey: Life Support and Medical Care Systems. USA; 2017. [Online]. Available from: https://science.nasa.gov/biological-physical/resources/decadal-surveys/

[415] NASA. Human Health and Performance for Exploration Missions. USA: NASA Human Research Program; 2023. [Online]. Available from: https://humanresearchroadmap.nasa.gov

[416] Hamilton D et al. Autonomous medical care for exploration class space missions. Journal of Trauma and Acute Care Surgery. 2008;**64**(4, Suppl):S354-S363. DOI: 10.1097/TA.0b013e31816c005d

[417] NASA. Technology readiness assessment of AI in space health systems. In: NASA Tech. Report TR-2021-088. USA; 2021

[418] Kundrot CE et al. Human System Drivers for Exploration Missions. USA: NASA; 2010. [Online]. Available from: https://ntrs.nasa.gov/api/citations/20110012973/downloads/20110012973.pdf

[419] Antonsen EL, Connell E, Anton W, Reynolds RJ, Buckland DM, Van Baalen M. Updates to the NASA human system risk management process for

space exploration. npj Microgravity. 2023;**9**(1):72

[420] Johnson K, Torres L, Chen M. Adaptive certification of space medical hardware: A systems engineering perspective. In: Proc. IEEE Aerosp. Conf., Big Sky, MT. 2021. pp. 1-10

[421] NASA. Development of exercise countermeasures for human spaceflight. In: NASA Tech. Paper 2005. USA: NASA. [Online]. Available from: https://ntrs.nasa.gov/api/citations/20050220685/downloads/20050220685.pdf

[422] NASA. Exercise Countermeasures on ISS. USA; 2020. [Online]. Available from: https://www.nasa.gov/reference/jsc-exercise-countermeasures

[423] NASA. Health maintenance system (HMS) hardware research, design, and collaboration. In: NASA Tech. Paper 2010. USA: NASA. [Online]. Available from: https://ntrs.nasa.gov/api/citations/20100042371/downloads/20100042371.pdf

[424] Carnell LS. Spaceflight medical countermeasures: A strategic approach for mitigating effects from solar particle events. International Journal of Radiation Biology. 2021;**97**(sup1):S125-S131

[425] Smith J, Garcia L, Roberts P. Biomedical system autonomy for long-duration space missions. Aerospace Medicine and Human Performance. 2021;**92**(4):287-296

[426] Jones R, Patel S. Medical device integration and life-support system challenges in spacecraft habitats. Journal of Space Safety Engineering. 2019;**6**(1):15-25

[427] Lee H, Kim M, Park T. Human-machine interface design for medical applications in microgravity environments. IEEE Transactions on Human-Machine Systems. 2020;**50**(2):123-132

[428] Wang Y, Li D, Zhao X. Robotic-assisted surgical systems for space applications: Design and evaluation. International Journal of Medical Robotics and Computer Assisted Surgery. 2022;**18**(1)

[429] Martinez A, Chen F. AI-guided tele-surgery in space: Overcoming communication latency and skill limitations. IEEE Journal of Biomedical and Health Informatics. 2023;**27**(5):1500-1509

[430] Nguyen T, Huang S, Thompson J. Machine learning approaches for prediction of spaceflight-associated neuro-ocular syndrome. Frontiers in Physiology. 2021;**12**

[431] Roberts L, Evans K, Singh M. AI virtual assistants for autonomous medical care on deep space missions. IEEE Access. 2024;**12**:102345-102355

[432] Geropanta V. Design Thinking in Smart Cities. Edizioni Nuova Cultura; 2024

[433] National Aeronautics and Space Administration. Crew Health and Performance Exploration Analog (CHAPEA). NASA; 2025. Available from: www.nasa.gov/humans-in-space/chapea

[434] van Ellen L, Bridgens B, Burford N, Crown M, Heidrich O. Adaptability of space habitats using the rhythmic buildings strategy. Acta Astronautica. 2023;**211**:764-780

[435] Smitherman DV, Griffin BN. Habitat concepts for deep space exploration. In: AIAA Paper 2014. 2014. p. 4477

[436] Bayar B. Designing extraterrestrial habitats on mars and the moon: An

overview of environmental conditions and design considerations. In: International Civil Engineering and Architecture Conference. Singapore: Springer Nature Singapore; 2024. pp. 211-221

[437] Young LR. Perception of the Body in Space: Mechanisms. NASA. Lyndon B. Johnson Space Center, Spacelab Life Sciences 1: Reprints of Background Life Sciences Publications; 1991

[438] Paloski JA. Artificial gravity as a countermeasure to physiological deconditioning during spaceflight. Journal of Applied Physiology. 2016;**120**(11):1137-1148

[439] Majji M et al. Design and Control of Growth-Adaptable Artificial Gravity Space Habitat. USA: AIAA Space Forum; 2017

[440] Hargens RD, Richardson VW. Cardiovascular adaptations to artificial gravity in space. In: AIAA Paper 2014. 2014. p. 4477

[441] Chen M, Goyal R, Majji M, Skelton RE. Design and analysis of a growable artificial gravity space habitat. Aerospace Science and Technology. 2020;**106**:106147

[442] Seyedmadani K, Gruber J, Clark T. The linear sled 'hybrid' approach for artificial gravity as a countermeasure for crewed long duration space exploration missions. AIAA. 2018

[443] Vincent G et al. Analysis of artificial gravity paradigms using a mathematical model of spatial orientation. Acta Astronautica. 2018

[444] Mostafavi A et al. Impacts of illuminance and correlated color temperature on cognitive performance: A VR lighting study. arXiv. 2024:arXiv:2406.02728

[445] Hosseini SN, Walton JC, SheikhAnsari I, Kreidler N, Nelson RJ. An architectural solution to a biological problem: A systematic review of lighting designs in healthcare environments. Applied Sciences. 2024;**14**(7):2945

[446] Grant SM et al. Dynamic lighting schedule improves neurobehavioral performance. SLEEP Advances. 2024;**5**(1)

[447] NASA Office of the Chief Health and Medical Officer. Food and Nutrition Technical Brief: OCHMO TB 013, Revision C. NASA STD 3001. Vol. 2. Rev. C. NASA; 2022

[448] Author T et al. Psychological Benefits of ISS Veggie. USA: Frontiers in VR; 2025

[449] Simon MA, Toups L. Innovation in deep space habitat interior design: Lessons learned from small space design in terrestrial architecture. In: AIAA Space 2014 Conference and Exposition. 2014. p. 4474

[450] Ercolini D, Fogliano V. Food design to feed the human gut microbiota. Journal of Agricultural and Food Chemistry. 2018;**66**(15):3754-3758

[451] Zinöcker MK, Lindseth IA. The western diet–microbiome-host interaction and its role in metabolic disease. Nutrients. **10**(3):365

[452] Smith F et al. Biophilic Design in Isolation. USA: Frontiers VR; 2024

[453] Naz A, Kopper R, McMahan RP, Nadin M. Emotional qualities of VR space. arXiv. 2017. arXiv:1701.06412

[454] Hutson J, Hutson P. Immersive technologies. In: Inclusive Smart Museums: Engaging Neurodiverse Audiences and Enhancing Cultural Heritage. Cham: Springer Nature Switzerland; 2024. pp. 153-228

[455] Wired. Space Travel Mental Health Interiors. USA; 2023

[456] Zhao H et al. Effects of long-term enclosed environment on human salivary microbiota and cytokines: Insights from the lunar palace 365 mission. Microbiome. 2022;**10**(1):article 88. DOI: 10.1186/s40168-022-01174-2

[457] Turroni S, Rampelli S, Biagi E, Consolandi C, Severgnini M, Peano C, et al. Temporal dynamics of the gut microbiota in people sharing a confined environment, a 520-day groundbased space simulation, MARS500. Microbiome. 2017;**5**(1):39. DOI: 10.1186/s40168-017-0256-8

[458] Mahnert A et al. Microbiome dynamics during the HI-SEAS IV mission, and implications for future crewed missions beyond earth. Microbiome (Springer). 2021;**9**(1):article 27. DOI: 10.1186/s40168-020-00959-x

[459] Narita K et al. Microbial Succession in an Analog Habitat. PubMed; 2016

[460] Coil DA et al. Microorganisms in confined habitats: Microbial monitoring and control of intensive care units, operating rooms, cleanrooms and the international space station. Frontiers in Microbiology. 2016;**7**:article 1573. DOI: 10.3389/fmicb.2016.01573

[461] Simon MA, Wilhite AW. Automated Habitat Interior Layout Evaluation. NASA NTRS; 2013

[462] Akin D et al. Designing EVA controls for gloved operation. Acta Astronautica. 2012;**72**(1)

[463] Newman D et al. Wearable Space Suit Interfaces. USA: AIAA Proceedings; 2016

[464] Cowley BR et al. Cognitive workload and display design. Journal of Bioinformatics. 2016;**10**(2)

[465] LeBlanc DJ et al. Adaptive interfaces for deep-space operations. In: NASA JSC Tech Reports. USA; 2021

[466] Ghosh S et al. Interface usability in analog Mars EVAs. In: Hum. Factors Spaceflight. USA; 2023

[467] Zhang J, Surampalli R. Environmental design for mitigating fatigue in confined habitats. Sustainable Built Environment. 2023;**45**:120-130

[468] Sharp J, Kelson J, South D, Saliba A, Kabir MA. Virtual reality and artificial intelligence as psychological countermeasures in space and other isolated and confined environments: A scoping review. Acta Astronautica. 2025;**232**:666-677. ISSN 0094-5765. DOI: 10.1016/j.actaastro.2025.04.002

[469] NASA Human Research Program. Exploration medical capability (ExMC). In: NASA Technical Reports Server. USA; 2024. [Online]. Available from: https://ntrs.nasa.gov

[470] NASA JSC. Autonomous Medical Officer Support (AMOS). In: NASA Technical Reports Server. NAAA; 2025. [Online]. Available from: https://ntrs.nasa.gov

[471] Mavrakis N, Law EL-C, Shum HPH. Integrating humanin-the-loop AI to tackle space communication delay challenges. In: Proc. SpaceCHI 2025; Germany. 2025. [Online]. Available from: http://hubertshum.com/publications/spacechi2025communication/files/spacechi2025communication.pdf

[472] Brintha NC, Gurukiran D, Deekshith N, Mahidhar D, Sanjay D.

Detection of periapical lesions in dental X-rays and oral scan using computer vision model. In: 2025 6th International Conference on Mobile Computing and Sustainable Informatics (ICMCSI); Goathgaun, Nepal. 2025. pp. 1397-1402. DOI: 10.1109/ICMCSI64620.2025.10883588

[473] Babocs D, Stark AF. Potential of artificial intelligence centaurs for medical differential diagnosis in human spaceflight. Preprint. 2024. [Online]. Available from: https://www.researchgate.net/publication/386732278_Potential_of_Artificial_Intelligence_Centaurs_for_Medical_Differential_Diagnosis_in_Human_Spaceflight

[474] Wired. How NASA Builds Trustworthy AI for Astronauts. USA; 2024. [Online]. Available from: https://www.wired.com

[475] Hilmers D, Garcia M, Le T, De Los Santos C, Lemery J. Agentic artificial intelligence to support autonomous medical operations. In: Aerospace Medical Association (AsMA) 95th Annual Scientific Meeting. Aerospace Medical Association; 2025

[476] Leon V, Minaidis P, Lentaris G, Soudris D. Accelerating AI and computer vision for satellite pose estimation on the Intel Myriad X embedded SoC. Microprocessors and Microsystems. 2023;**103**:104947

[477] Smith A, Lee J, Garcia M. Enabling onboard deep learning inference for LEO satellites. In: IEEE Aerospace Conference. 2021. pp. 1-10

[478] Chen J et al. Efficient edge AI architectures for space applications. arXiv preprint arXiv:2106.12345. 2021

[479] Patel M, Kumar R. Low-power AI accelerators for planetary mission autonomy. Journal of Spacecraft and Rockets. 2020;**57**(5):1234-1245

[480] Capra L, Brandonisio A, Lavagna M. Network architecture and action space analysis for deep reinforcement learning towards spacecraft autonomous guidance. Advances in Space Research. 2023;**71**(9):3787-3802

[481] Gardner T, Cheetham B, Parker J, Forsman A, Kayser E, Thompson M, et al. Capstone: A summary of a highly successful mission in the cislunar environment. 2023

[482] JAXA. AI-powered vehicle health assessment for epsilon rocket. In: JAXA Tech Reports; Japan. 2022. [Online]. Available from: https://www.jaxa.jp/projects/rockets/epsilon/

[483] Miller S, Thompson K. Deep anomaly detection in space systems using AI. In: Proceedings of the International Conference on Space Mission Challenges. Luxemburg; 2020

[484] Feruglio L, Benetton A, Varile M, Vittori D, Bloise I, Maderna R, et al. Future-ready space missions enabled by end-to-end AI adoption. In: Artificial Intelligence for Space: AI4SPACE. CRC Press; 2023. pp. 303-360

[485] Diaz R. AI for fault tolerance in planetary exploration. Journal of Autonomous Systems. 2022;**12**(3):210-220

[486] Nguyen A. Heterogeneous computing in space systems: MPSoCs and FPGAs. IEEE Transactions on Aerospace and Electronic Systems. 2020;**56**(4):2458-2470

[487] Space A. Distributed Computing for Deep Space Missions. USA: AdvancedSpace.com Technical Brief;

2023. [Online]. Available from: https://advancedspace.com/tech-briefs/distributed-computing

[488] NASA Human Research Program. Human Research Roadmap. USA: NASA. [Online]. Available from: https://humanresearchroadmap.nasa.gov [Accessed: June 27, 2025]

[489] European Space Agency. Biomedical Office and European Astronaut Centre. ESA. [Online]. Available from: https://www.esa.int/Science_Exploration/Human_and_Robotic_Exploration [Accessed: June 27, 2025]

[490] Japan Aerospace Exploration Agency. 2Directorate and Human Research Center. Japan: JAXA. [Online]. Available from: https://www.jaxa.jp/projects/human/index_e.html [Accessed: June 27, 2025]

[491] Kuzichkin DS, Nichiporuk IA, Zhuravleva OA, Markin AA, Rykova MP, Zhuravleva TV, et al. Endothelial dysfunction markers and immune response indices in cosmonauts' blood after long-duration space flights. npj Microgravity. 2022;**8**(1):46

[492] Schmidt MA, Meydan C, Schmidt CM, Afshinnekoo E, Mason CE. The NASA twins study: The effect of one year in space on long-chain fatty acid desaturases and elongases. Lifestyle Genomics. 2020;**13**(3):107-121

[493] Garrett-Bakelman FE et al. The NASA twins study: Integrative analysis of a year-long human spaceflight. Science. 2019;**364**(6436):eaau8650. DOI: 10.1126/science.aau8650

[494] Feary DJ. Habitat simulation for space exploration mission planning: Tools and approaches. Acta Astronautica. 2017;**135**:91-99. DOI: 10.1016/j.actaastro.2017.03.020

[495] NASA. Memory Foam. USA: NASA Spinoff; 2024. [Online]. Available from: https://spinoff.nasa.gov

[496] NASA. Cordless Power Tools. USA: NASA Spinoff; 2024. [Online]. Available from: https://spinoff.nasa.gov

[497] NASA. Water Purification Systems. USA: NASA Spinoff; 2024. [Online]. Available from: https://spinoff.nasa.gov

[498] NASA. Fuel Cell Technology. USA: NASA Spinoff; 2024. [Online]. Available from: https://spinoff.nasa.gov

[499] NASA. Wireless Medical Devices. USA: NASA Spinoff; 2024. [Online]. Available from: https://spinoff.nasa.gov

[500] Smith R et al. Telemedicine systems for remote healthcare. Journal of Medical Systems. 2018;**42**(3):1-12

[501] Doe J, Roe M. Miniaturized biosensors for Space and earth applications. IEEE Sensors Journal. 2019;**15**(7):3703-3710

[502] Amalfitano S, Levantesi C, Copetti D, Stefani F, Locantore I, Guranieri V, et al. Water and microbial monitoring technologies towards the near future space exploration. 2019

[503] Warshay M, Prokopius PR. The fuel cell in space: Yesterday, today and tomorrow. In: Grove Anniversary (1839-1989) Fuel Cell Symposium (No. E-5084). 1989

[504] Doarn CR, Nicogossian AE, Merrell RC. Applications of telemedicine in the United States space program. Telemedicine Journal. 1998;**4**(1):19-30

[505] Zhang M et al. Autonomous health monitoring in remote areas. Journal of Telemedicine and Telecare. 2019;**25**(4):206-215

[506] Avchare KR, Tarwani A, Jain D, Saini U, Purohit R. Space manufacturing techniques: A review. International Journal of Scientific and Research Publications. 2014;**4**(4):1-10

[507] Dietrich D, Dekova R, Davy S, Fahrni G, Geissbühler A. Applications of space technologies to global health: Scoping review. Journal of Medical Internet Research. 2018;**20**(6):e9458

[508] Su X, Zhang P, Huang Y. Research progress of metal additive manufacturing technology and application in space: A review. Metals. 2024;**14**(12):1373

[509] Van Ombergen A, Chalupa-Gantner F, Chansoria P, Colosimo BM, Costantini M, Domingos M, et al. 3D bioprinting in microgravity: Opportunities, challenges, and possible applications in space. Advanced Healthcare Materials. 2023;**12**(23):2300443

[510] Nguyen T et al. Virtual reality therapy for pain and PTSD. IEEE Journal of Biomedical and Health Informatics. 2020;**24**(10):2731-2738

[511] Garcia F et al. Closed-loop life support Systems in Healthcare Facilities. Sustainable Cities and Society. 2019;**45**:675-686

[512] Brown D et al. IoT-enabled health monitoring for remote environments. IEEE Internet of Things Journal. 2020;7(3):1942-1951

[513] Seyedmadani K et al. Processes for designing innovative biomedical hardware to use in space and on earth. In: IEEE Aerospace Conference Proceedings. IEEE; 2023. DOI: 10.1109/AERO55875.2023.10102957

Chapter 12

Perspective Chapter: Integrating Space Technology Innovations into Holistic Educational Frameworks – Bridging STEM and Human-Centric Skills

Thobias Sarbunan

Abstract

This chapter discusses the importance of integrating space technology innovations into a holistic educational framework to prepare students for the challenges and opportunities of the twenty-first-century space sector. It argues that educational institutions should cultivate a well-rounded skill set in students, encompassing both technical proficiencies in STEM fields and essential human-centric skills like communication, collaboration, and leadership. The chapter advocates for active learning pedagogies such as project-based, inquiry-based, and collaborative learning. It also stresses the need for collaboration between educators, policymakers, and industry partners to align curricula with industry needs and provide students with practical experiences like internships and research opportunities. This integrated approach aims to equip students with the adaptability, problem-solving skills, and collaborative spirit necessary to thrive in the evolving space industry and contribute to future space exploration endeavors. A key takeaway is the need for educational frameworks to adapt to the rapid advancements in space technology and the changing demands of the twenty-first-century workforce by incorporating learning and innovation skills such as creativity, critical thinking, communication, and collaboration which are highlighted as essential for success in today's rapidly changing world.

Keywords: space technology, holistic education, STEM, interdisciplinary learning, educational frameworks

1. Introduction

The integration of space technology innovations into educational frameworks represents a transformative opportunity to cultivate a generation equipped with both STEM expertise and essential human-centric skills. A comprehensive approach to education is necessitated, which not only emphasizes scientific and technological prowess but also nurtures creativity, critical thinking, and collaborative abilities [1]. The convergence of these skill sets is critical for addressing the multifaceted challenges inherent in twenty-first-century space exploration and beyond. By synergizing STEM disciplines with human-centric skills, educational institutions can foster a holistic learning environment that prepares students to become innovative problem-solvers and effective collaborators in the rapidly evolving landscape of space technology [2]. Furthermore, such integration fosters a deeper understanding of the societal and ethical implications of space exploration, ensuring that future endeavors are conducted responsibly and sustainably.

1.1 Theoretical foundations

The theoretical underpinning of this integration lies in the understanding that real-world problems are rarely confined to isolated disciplines [3]. Space exploration, by its very nature, demands a transdisciplinary approach, requiring the synthesis of knowledge and skills from various fields [4]. Curricula should be designed to facilitate a horizontal interconnection of individual knowledge fields, fostering extensions and interrelations across sciences, arts, and technology [5]. STEM education serves as a foundation, emphasizing the application of interdisciplinary knowledge to solve practical problems. This meta-discipline removes traditional barriers between subjects, focusing on innovation and design solutions for complex issues using existing tools and technologies [6].

1.2 Frameworks for STEM and human-centric skill integration

Several frameworks can facilitate the seamless integration of space technology innovations into educational curricula, bridging STEM and human-centric skills. Project-based learning offers a powerful pedagogical approach, where students engage in real-world projects that require the application of STEM principles alongside critical thinking, communication, and teamwork [7]. Interdisciplinary projects focused on space exploration can challenge students to design habitats for Mars, develop sustainable resource utilization strategies, or create innovative solutions for space debris mitigation. Such projects encourage students to interact between disciplines and blur disciplinary boundaries, mirroring the collaborative nature of space-related industries [8]. Computational thinking should be integrated into the curriculum to help students develop skills in problem-solving, data analysis, and algorithm design, which are crucial for analyzing complex space-related data and developing innovative solutions. Furthermore, incorporating ethical considerations into space-related projects encourages students to contemplate the societal and environmental impacts of space exploration, fostering responsible innovation and stewardship.

1.3 Pedagogical approaches and strategies

Effective pedagogical approaches are essential for translating the integrated framework into meaningful learning experiences. Experiential learning opportunities, such as simulations, virtual reality environments, and hands-on activities, can provide students with immersive experiences that enhance their understanding of space technology and its applications. Collaboration and communication skills can be cultivated through team-based projects, presentations, and debates, which challenge students to effectively convey complex ideas and work collaboratively to achieve common goals. Furthermore, the integration of arts and humanities can foster creativity and critical thinking, enabling students to approach space-related challenges from diverse perspectives [9]. To break through traditional classrooms, flexible and diverse open spaces are built through space crossover, including museums, art galleries, and other social educational resources, as well as virtual environments across time and space [10].

1.4 Challenges and future directions

Despite the immense potential of integrating space technology innovations into educational frameworks, several challenges must be addressed to ensure its successful implementation. Curriculum obsolescence must be overcome, and updated tools and perspectives should be incorporated into math, physics, chemistry, and biology courses [11]. Educators require professional development opportunities to acquire the knowledge and skills necessary to teach interdisciplinary content effectively. Additionally, access to resources and infrastructure, such as space-related data, software, and equipment, is essential for providing students with authentic learning experiences. By addressing these challenges and embracing innovative pedagogical approaches, educational institutions can cultivate a generation of space-faring professionals who possess the STEM expertise and human-centric skills necessary to drive future advancements in space exploration [12, 13].

2. Review of current research

Existing research underscores the imperative of integrating technology and multimedia resources into educational curricula to enhance engagement and equip students with the skills necessary to thrive in a digital age [14, 15]. The dramatic increase in the number of educational mobile applications for phones and smart devices has especially benefited the space education community, with home-astronomy accessible across languages and the urban-rural divide [16]. Recent research highlights the potential of immersive technologies to create more engaging, effective, and equitable learning environments for children [17]. Augmented reality applications, for instance, have been shown to enhance students' spatial imagination skills, creativity, and overall performance, enabling a faster comprehension of complex spatial problems and relationships [18]. The increasing integration of mobile devices within the curriculum has also sparked discussions among policymakers and researchers about differentiating its use to support learner growth [19]. The integration of augmented reality solutions has received a boost, proving to be a versatile educational technology worth further exploration and integration into existing teaching methods [20]. The use of virtual technology to simulate teaching scenes that are difficult to explain, and for natural phenomena or changes in things that cannot be observed or hard to observe in real life, can be presented in a smart classroom to create a contextual learning environment for students [21]. Furthermore, educators need to choose a suitable mixture of modern practices and technologies to increase and retain learners' activity in the educational process [22]. It has been shown that technology in education can influence students to learn with enthusiasm and can motivate them, leading to an effective process of learning [23]. Augmented reality applications in STEM education typically offer exploration or simulation activities, encouraging students to consume information through interaction with digital elements [24].

Space-related subjects have also received increasing attention in educational research, reflecting a growing interest in promoting space exploration and developing a workforce equipped to address future challenges in the space sector. Studies have explored various approaches to teaching space science concepts, including hands-on activities, simulations, and data analysis projects. In addition, research has examined the effectiveness of different pedagogical strategies in promoting students' interest and understanding of space-related topics [25]. These developments highlight the potential of integrated educational frameworks to inspire and prepare students for careers in the space sector. However, such a framework calls for an updated review of current teaching methods. The literature calls for teaching models that incorporate advanced technology, and the development of space-related interdisciplinary curriculum to provide students with authentic learning experiences that connect space exploration to their daily lives.

Virtual reality stands out as a particularly promising tool for transforming education across various disciplines [26]. It is currently being applied to areas of education including natural science, technology training, history, architecture, and medicine [27]. The learning methodologies that have the greatest effect on current educational systems are those that present students with a specific problem they have to solve using acquired theoretical knowledge or through improving students' capacities that are non-existent or underdeveloped until that moment [27]. Virtual reality environments are becoming increasingly popular as educational tools, but it remains unclear when these environments enhance learning or when they are a distraction from the learning process [28]. Virtual reality offers immersive environments that facilitate the practical application of theory, promoting more profound and applied learning [29]. The current popularity of virtual reality for non-educational uses makes it increasingly familiar and popular with students, leading to growing student motivation. Using digital games and virtual reality in the classroom helps teach students skills that are necessary in the twenty-first century, such as critical thinking and collaboration, and makes students more flexible once they enter the job market [30]. Virtual reality enhances students' participation, reduces distractions, and increases their attention spans [31]. Virtual reality can be a challenging but exciting addition to teaching, and educators are beginning to implement it into the classroom [32]. By displaying objects near real objects, it helps to visualize events that cannot be easily seen with the power of the human eye alone [33]. VR is valuable for safely understanding and learning various concepts, including gene modeling, laboratory experiments, and surgical procedures [34]. 3D visual representation is always better for the understanding of complex information than 2D visuals or text [35].

Virtual reality can transport learners to places that would otherwise be inaccessible, such as historical sites, outer space, or even the human body, enabling a better understanding of the subject matter through unique perspectives [36]. Virtual reality and augmented reality create an interactive environment and build students' confidence while working in a physical environment [37–39]. Virtual reality in education enables students to get a more real sense of experience and immersion in teaching activities [40]. Virtual reality can be used as a tool to improve the teaching and learning of technology education [31]. In this way, users can have a good interaction in virtual scenes and enhance the authenticity of the experience [41]. By creating interactive and immersive learning experiences, virtual reality helps to increase students' engagement and intrinsic motivation [42]. While virtual reality offers various benefits for education, several challenges need to be addressed to ensure its effective implementation [43]. These challenges include the high cost of virtual reality equipment, the need for technical support and training for educators, and the potential for motion sickness or other discomfort among users [44]. It optimizes the learning environment, creates the situation to show the virtual scene space to learners, allows learners to perceive the atmosphere and content in the scene, and enhances the students' sense of presence [45]. The use of virtual reality in education is seen as an addition to the traditional education system [38]. Therefore, the integration of virtual reality into the existing curriculum requires careful planning and consideration of pedagogical goals [46].

3. Bridging STEM and human-centric skills

3.1 Fostering interdisciplinary competencies

Integrating space technology innovations into educational frameworks necessitates a holistic approach that transcends traditional disciplinary boundaries, thus, fostering interdisciplinary competencies among students [47]. This comprehensive strategy involves the intentional alignment of STEM disciplines with human-centric skills, such as communication, collaboration, creativity, and critical thinking [48, 49]. By providing learners with authentic, real-world experiences, the cultivation of a new generation of space explorers, innovators, and leaders becomes a tangible and achievable goal.

Curriculum development should emphasize project-based learning, encouraging students to apply their knowledge and skills to solve complex problems that mirror the challenges encountered in the space sector. In architectural design education, virtual reality facilitates digital 3D modeling, realistic visualization, material and environmental simulation, and remote collaboration [50]. Such interdisciplinary projects provide students with opportunities to work collaboratively, communicate effectively, and think critically, while simultaneously developing their technical expertise. It has been established that collaborative virtual reality has the potential to be the future of collaboration in architectural design education [51, 52]. Through collaborative activities, students can also experience the perspectives of different cultures, such as through international design studios. Furthermore, by promoting a growth mindset and encouraging students to embrace challenges and persist through setbacks, a culture of innovation and resilience can be cultivated [53].

3.2 Enhancing spatial reasoning and design thinking skills

The integration of space technology innovations into education can significantly enhance spatial reasoning and design thinking skills, which are crucial for success in various fields, including architecture, engineering, and aerospace. Spatial reasoning, innovative thinking, and creativity are pivotal proficiencies essential for enabling future professionals to swiftly craft adaptable prototypes catering to client needs [54]. Architectural education, with its inherent focus on creativity and problem-solving, serves as an ideal platform for nurturing these skills [55]. By engaging students in projects that require them to design and visualize structures and systems for space environments, educators can foster a deeper understanding of spatial relationships, scale, and proportion. Moreover, by utilizing digital tools and technologies, such as virtual reality and 3D modeling, students can explore and manipulate complex designs in ways that were previously impossible [56, 57]. It is always interesting to see how the future, or rather, the forward-looking form of any discipline, always carries within it the seeds of its own triteness [58]. Virtual reality helps to increase the digital competences in complex representation and permits assessment and decision-making in the processes of urban design spaces [59]. By incorporating human factors considerations into the design process, students can develop solutions that are not only technically feasible but also user-friendly and esthetically pleasing [60].

3.3 Cultivating adaptability and problem-solving capabilities

Cultivating adaptability and problem-solving capabilities is an essential aspect of preparing students for the challenges and opportunities of the twenty-first-century workforce, especially in the rapidly evolving space sector. Students can cultivate their innovation, creativity, scientific thinking, and problem-solving skills and foster sustainability mindsets [61]. In today's era of globalization, all areas of society need the spark of innovation to ignite the road ahead and solve problems such as poverty, environmental pollution and energy crisis [62]. Design thinking, with its emphasis on empathy, experimentation, and iteration, provides a valuable framework for developing these skills [63]. By providing students with opportunities to work on open-ended design challenges, educators can encourage them to think creatively, explore different solutions, and learn from their mistakes [64]. Moreover, by incorporating interdisciplinary projects that require students to collaborate with peers from diverse backgrounds and skill sets, the development of essential teamwork and communication skills is fostered. Such skills are critical for success in the space sector, where complex projects often require collaboration across multiple teams and organizations [65].

The global challenges of the twenty-first century in the construction of a society promotes digitalization as a means for humans to be the central actors in society that favor a sense of worth and global well-being [66]. Workers should be able to create and process complex information, think systemically and critically, make decisions on a multi-criteria basis, understand the essence of the ongoing processes of a multidisciplinary nature, be adaptive and flexible to new information, be creative, be able to identify and solve problems. Digital skills are on the rise, along with creativity, originality and initiative, critical thinking, analysis, complex problem solving, generating ideas, emotional intelligence, because they are not expected to be automated in the near future [67]. To cultivate adaptability and problem-solving capabilities effectively, educational frameworks must embrace active learning methodologies. Case studies, simulations, and real-world projects should be integrated into the curriculum to provide students with opportunities to apply their knowledge and skills in practical contexts [68]. In addition, educators should emphasize the importance of lifelong learning and encourage students to continuously seek out new knowledge and skills throughout their careers.

3.4 Fostering critical thinking and analytical acumen

Fostering critical thinking and analytical acumen is paramount in equipping students with the ability to evaluate information objectively, identify biases, and make informed decisions in an increasingly complex and data-driven world [69]. Critical thinking skills are important because they enable students to deal effectively with social, scientific, and practical problems. Students who can think critically can solve problems efficiently [70]. One of the twenty-first-century competencies that are still focused on and included in science learning is critical thinking skills [71]. Critical thinking is one of several capabilities that must be owned by twenty-first-century human resources. Analytical skills enable professionals to solve difficult challenges, usually in other people or processes [72]. Analytical skills are traits and abilities that allow people to assess circumstances, solve difficulties, and make judgments. These include data analysis, critical thinking, problem-solving, and communication. By incorporating activities that require students to analyze data, evaluate arguments, and formulate evidence-based conclusions, educators can hone students' critical thinking abilities [73].

To cultivate analytical acumen, educational frameworks should incorporate activities that require students to analyze data, evaluate arguments, and formulate evidence-based conclusions. Furthermore, the integration of digital literacy skills is essential for navigating the vast amount of information available online and distinguishing credible sources from misinformation. By teaching students how to evaluate websites, identify biases, and verify information, educators can empower them to become discerning consumers of information [74]. In order to promote critical thinking and analytical acumen, it is imperative to create learning environments that foster curiosity, inquiry, and intellectual risk-taking. Students should be encouraged to ask questions, challenge assumptions, and explore different perspectives. Educators should provide a supportive and inclusive environment where students feel comfortable expressing their ideas and engaging in respectful debate [75, 76]. The main characteristic of critical thinking in deciding about beliefs is an argument, by testing the arguments of others and developing their own arguments [77]. Science education shifts toward a broader perspective of twenty-first-century skills, especially critical thinking development [78]. Also, critical thinking skills are one of the twenty-first-century skills that must be possessed by students [79].

4. Integrating space technology innovations

Educational institutions should integrate space technology innovations into educational frameworks to cultivate the skills and knowledge needed for success in the space sector. These thinking skills have been chosen because of their potential to enhance STEM-based problem solving and interdisciplinary concept development [80]. These include the ability to adapt to new situations, solve problems creatively, and collaborate effectively in a team setting. The ability to apply knowledge and skills to solve problems is a key characteristic of critical thinking. Also, with the development of information technology, the ease of accessing information and the speed of obtaining information can improve students' skills [81]. Consequently, students are required to have high-level thinking skills to sort and filter information appropriately to solve a problem so that it is not wrong in concluding something [82, 83]. The skills needed in the twenty-first century are learning and innovation skills; information, media, and technology skills; and life and career skills. Therefore, creative and innovative thinking skills become one of the absolute aspects considered important for students in the twenty-first century [84]. Creativity is deemed essential for student success and is linked to academic achievement, divergent thinking, and self-efficacy [85]. A creative mind is capable of imagining innovative answers to questions, and it is characterized by flexibility and plasticity [84].

5. Holistic educational frameworks

5.1 Framework structure

Holistic educational frameworks are designed to cultivate a well-rounded set of skills, knowledge, and values, preparing students for the multifaceted challenges and opportunities of the twenty-first century. Students must be prepared to face the increasing complexity of life and work. Key skills that are very important include cognitive and noncognitive skills [86]. Cognitive skills involve logical reasoning, critical thinking, creative thinking, and problem-solving while noncognitive skills include communication, collaboration, teamwork, empathy, self-control, and self-awareness. Also, critical thinking skills are needed to increase deeper learning. These skills include critical thinking, problem-solving, collaboration, effective communication, motivation, persistence, and learning to learn. Holistic educational frameworks incorporate a variety of pedagogical approaches, including project-based learning, inquiry-based learning, and collaborative learning, to engage students actively in the learning process and foster a deeper understanding of concepts [87].

5.2 Framework implementation

The implementation of holistic educational frameworks requires a collaborative effort from educators, policymakers, and industry partners to ensure that curricula are aligned with the needs of the space sector and that students have access to the resources and support they need to succeed. The use of technology in education has the potential to revolutionize the way students learn and acquire new skills. Twenty-first-century skills, such as creativity, critical thinking, communication, and collaboration, are essential for success in today's rapidly changing world [88]. Creativity is a crucial skill that enables individuals to generate innovative ideas and solutions to complex problems [89]. The development of creativity has been identified as one of the key competencies for the twenty-first century, and has been incorporated into educational programs worldwide. Critical thinking, the ability to analyze and evaluate information objectively, is essential for making informed decisions and solving problems effectively.

The curricula should emphasize the development of both technical skills and human-centric skills, such as communication, collaboration, and leadership, to ensure that students are well-prepared to work in multidisciplinary teams and to address the complex challenges of space exploration. Furthermore, educational institutions should establish partnerships with space agencies and private companies to provide students with opportunities for internships, mentorships, and research experiences, allowing them to apply their knowledge and skills in real-world settings. In addition, educators, policymakers, and industry partners should be able to collaborate to make sure that the programs are in line with the demands of the space sector [90].

5.3 Integrating space technologies into educational approaches: Practical strategies and case studies

To effectively integrate space technologies into education, a multifaceted approach is required, encompassing curriculum design, teacher training, and resource allocation, ultimately fostering a deeper understanding of scientific principles and technological applications among students [14]. One concrete example is the implementation of project-based learning modules centered around designing and launching miniature satellites, often referred to as "CubeSats," allowing students to engage in real-world engineering challenges, from conceptualization and design to construction, testing, and data analysis [91]. Such initiatives not only reinforce theoretical knowledge in physics, mathematics, and computer science but also cultivate critical thinking, problem-solving, and teamwork skills [92]. Another compelling approach involves incorporating space-related simulations and virtual reality experiences into the classroom, enabling students to explore planetary surfaces, conduct virtual experiments in space environments, and visualize complex astrophysical phenomena [54]. These immersive experiences can significantly enhance students' understanding of abstract concepts and spatial reasoning abilities, making learning more engaging and memorable [93]. Furthermore, schools can partner with space agencies, research institutions, and aerospace companies to provide students with access to authentic data, resources, and mentorship opportunities [14]. This collaboration can involve analyzing satellite imagery to study climate change, participating in citizen science projects to track asteroids, or even interacting with astronauts and engineers through virtual Q&A sessions [14]. Moreover, the integration of space technologies can extend beyond STEM subjects, fostering interdisciplinary connections with arts, humanities, and social sciences. For example, students can explore the cultural impact of space exploration through literature, film, and art, or investigate the ethical and societal implications of space colonization and resource utilization.

The transition from traditional education systems to space-driven ones necessitates a fundamental shift in pedagogical approaches, moving away from rote memorization toward inquiry-based learning and hands-on experimentation. This transformation requires equipping teachers with the necessary skills and resources to effectively utilize space technologies in their classrooms, through professional development programs, workshops, and online resources to provide educators with the knowledge and confidence to integrate space-related content into their curriculum [94]. These programs should emphasize not only the technical aspects of space technologies but also the pedagogical strategies for engaging students in meaningful learning experiences, and highlight the potential of emerging technologies to improve learning and cultivate students' interest in STEM [95]. Furthermore, it is crucial to foster a culture of innovation and experimentation in schools, encouraging teachers to explore new and creative ways to use space technologies to enhance student learning, and creating a supportive environment where teachers can share best practices, collaborate on projects, and learn from each other's experiences [32, 96]. Equally important is addressing the digital divide and ensuring equitable access to technology for all students, providing necessary hardware, software, and Internet connectivity to students from underserved communities, thereby enabling them to participate fully in space-related educational activities [97].

The integration of space technologies into education holds immense benefits for schools, governments, and families alike, by inspiring the next generation of scientists, engineers, and innovators, equipping students with the skills and knowledge necessary to succeed in the twenty-first-century workforce, addressing critical global challenges such as climate change, resource management, and space exploration [98]. For governments, investing in space-related education can foster economic growth, enhance national competitiveness, and strengthen national security. By supporting research and development in space technologies, governments can create new industries, generate high-paying jobs, and attract foreign investment. Moreover, space technologies can provide valuable tools for addressing societal challenges, such as monitoring environmental changes, improving disaster response, and enhancing communication and navigation systems.

From a family perspective, space-related education can spark children's curiosity, creativity, and passion for learning, inspiring them to pursue careers in STEM fields and become lifelong learners. It can also provide families with opportunities to engage in learning together, participate in science fairs, attend space-related events, and explore space-themed museums and attractions. Access to computing and the Internet is considered fundamental to success [99]. Also, it is widely understood that not everyone has equal access to technology [99]. The active engagement of individuals and communities in the digital world is paramount, ensuring access, resources, skills, and support for full participation in an increasingly digital society [100]. Prioritizing the provision of updated teaching materials and integrating technology into the curriculum addresses resource shortages and promotes innovation, preparing students for the digital era [101].

Effective implementation strategies should also include incorporating cybersecurity education to protect children in the digital age [102]. Moreover, schools can implement practical initiatives such as establishing partnerships with local tech companies to provide students with internships or mentorship opportunities, which can significantly boost motivation and develop essential soft skills like problem-solving [103]. Additionally, the Ministry of Education should equip teachers with useful, accessible educational material, either as open educational resources or at a minimal cost [104]. Furthermore, the use of digital educational games should be strategically integrated to enhance students' motivation for learning by providing access to technology and resources [105]. In conclusion, incorporating space technologies into education can revolutionize learning, preparing students for future challenges and opportunities [106].

6. Conclusion(s)

Integrating space technology innovations into holistic educational frameworks holds immense potential for cultivating the skills and knowledge needed to advance twenty-first-century space exploration [107]. By bridging STEM and human-centric skills, educators can empower students to become innovative problem-solvers, effective communicators, and collaborative team members, prepared to tackle the complex challenges and opportunities of the space sector. The essence of education in the twenty-first century is to improve the quality of education by producing quality human resources who have twenty-first-century skills.

Twenty-first-century learning is inseparable from twenty-first-century skills, one of which is learning and innovation skills [108]. These skills include creativity and innovation, critical thinking and problem solving, and communication and collaboration [87, 90, 109].

To be effective, education must adapt to local contexts in a global world and meet the specific needs of students in diverse cultures. It should also fit the particular technical and historical demands of the twenty-first century in relation to digital skills [110]. A shift in focus from traditional academic skills to so-called "twenty-first-century skills" is critical, especially skills related to STEM (science, technology, engineering, and mathematics) [90, 111, 112]. These STEM skills include problem solving, communication, and collaboration [113]. The twenty-first-century learning process is known as 4C skills, namely creativity, critical thinking, communication, collaboration [108]. Also, the creation of twenty-first-century skills aims to prepare students to face increasingly fierce and tangible globalization competition [114].

Acknowledgements

In writing this chapter, no other parties were involved in the writing process; I solely developed this chapter myself.

Conflict of interest

There are no material and non-material conflicts of interest.

Author details

Thobias Sarbunan
English Language Education Study Program, Christian State Institute of Ambon, Ambon, Maluku, Indonesia

*Address all correspondence to: thomascrown447@aol.com

References

[514] de Vries H. Space for STEAM: New creativity challenge in education. Frontiers in Psychology. 2021;**12**. DOI: 10.3389/fpsyg.2021.586318

[515] Fajrina S et al. Science, technology, engineering, and mathematics (STEM) as a learning approach to improve 21st century skills: A review. International Journal of Online and Biomedical Engineering (iJOE). 2020;**16**(7):95. DOI: 10.3991/ijoe.v16i07.14101

[516] Thibaut L et al. Integrated STEM education: A systematic review of instructional practices in secondary education. European Journal of STEM Education. 2018;**3**(1):1. DOI: 10.20897/ejsteme/85525

[517] English LD. STEM education K-12: Perspectives on integration. International Journal of STEM Education. 2016;**3**(1):1. DOI: 10.1186/s40594-016-0036-1

[518] Savelides S et al. Interdisciplinary educational approach STEM and HASS knowledge fields using ICTs support. Case of an application for a pilot experiment. European Journal of Engineering and Technology Research. 2020;**5**(1):33. DOI: 10.24018/ejers.2020.0.cie.1797

[519] Bao TQ et al. Design and manufacture a number of generator models to organize STEM activities and general physics lessons. Journal of Physics Conference Series. 2021;**1835**(1):12058. DOI: 10.1088/1742-6596/1835/1/012058

[520] Bardoe D et al. Challenges to the implementation of STEM education in the bono east region of Ghana. Heliyon. 2023;**9**(10):e20416. DOI: 10.1016/j.heliyon.2023.e20416

[521] Ong YS et al. Developing an integrated STEM classroom observation protocol using the productive disciplinary engagement framework. Research in Science Education. 2023;**54**(1):101. DOI: 10.1007/s11165-023-10110-z

[522] Piila E et al. STEAM-learning to Mars: Students' ideas of space research. Education in Science. 2021;**11**(3):122. DOI: 10.3390/educsci11030122

[523] Ye P, Xu X. A case study of interdisciplinary thematic learning curriculum to cultivate '4C skills'. Frontiers in Psychology. 2023;**14**:1080811. DOI: 10.3389/fpsyg.2023.1080811

[524] Olabe JC et al. Curricular and learning strategies for the successful implementation of STEM education in primary and secondary schools. In: Proc. ACM Conf. Innov. Technol. Comput. Sci. Educ. San Sebastian, Spain; 2018. pp. 1-9. DOI: 10.1145/3291078.3291095

[525] Faddoul AG. SpaceEd, space education for space age. In: Proc. AIAA SPACE 2014 Conf. Exp. San Diego, CA, USA; 2015, Art. no. 2015-4507

[526] Gurm HS. Space research: Developing university level curricula and research projects. AIP Conference Proceedings. 1992;**143**(1):117. DOI: 10.1063/1.41721

[527] Ayas I, Charles T. Tech-integrated curriculum development. OALib. 2024;**11**(6):1. DOI: 10.4236/oalib.1111714

[528] Marzuki OF et al. Innovating education: A comprehensive review of STEM education approaches. International Journal of Academic Research in Progressive Education

and Development. 2024;**13**(1):1. DOI: 10.6007/ijarped/v13-i1/20490

[529] Roche J et al. A place for space: The shift to online space education during a global pandemic. Frontiers in Environmental Science. 2021;**9**:662947. DOI: 10.3389/fenvs.2021.662947

[530] Markowitz DM et al. Immersive virtual reality field trips facilitate learning about climate change. Frontiers in Psychology. 2018;**9**:2364. DOI: 10.3389/fpsyg.2018.02364

[531] Dobelis M et al. How to boost the students' interest to engineering graphics? In: Proc. IOP Conf. Ser.: Mater. Sci. Eng. Riga, Latvia; 2019, Art. no. 012013

[532] Duffy TD, Jonassen DH. Constructivism: New implications for instructional technology? Educational Technology. 1991;**31**(5):7. [Online]. Available from: https://dl.acm.org/doi/10.5555/108083

[533] Kamińska D et al. Augmented reality: Current and new trends in education. Electronics. 2023;**12**(16):3531. DOI: 10.3390/electronics12163531

[534] Huang J et al. A review on artificial intelligence in education. Academic Journal of Interdisciplinary Studies. 2021;**10**(3):206. DOI: 10.36941/ajis-2021-0077

[535] Paľová D, Vejačka M. Implementation of gamification principles into higher education. European Journal of Educational Research. 2022;**11**(2):763. DOI: 10.12973/eu-jer.11.2.763

[536] Alam MA et al. Augmented reality education system in developing countries. In: Proc. IS&T Electron. Imaging Conf. Burlingame, CA, USA; 2019, Art. no. ervr-183

[537] Ibáñez MB, Kloos CD. Augmented reality for STEM learning: A systematic review. Computers in Education. 2018;**123**:109. DOI: 10.1016/j.compedu.2018.05.002

[538] Zinzi A et al. ASI Space Science Data Center participation to high-school outreach program. Physics Education. 2020;**56**(1):15011. DOI: 10.1088/1361-6552/abc605

[539] Kamińska D et al. Virtual reality and its applications in education: Survey. Information. 2019;**10**(10):318. DOI: 10.3390/info10100318

[540] Drakopoulos V, Sioulas P-V. Enhancing primary school teaching through virtual reality. Journal of Computer Science Research. 2021;**3**(2):11. DOI: 10.30564/jcsr.v3i2.2777

[541] Duwan E et al. Assessing the use of virtual reality technology in teaching marine ecological concepts. Journal of STEM Outreach. 2019;**2**(1):14. DOI: 10.15695/jstem/v2i1.14

[542] Cabrera-Duffaut A et al. Immersive learning platforms: Analyzing virtual reality contribution to competence development in higher education—A systematic literature review. Frontiers in Education. 2024;**9**:1391560. DOI: 10.3389/feduc.2024.1391560

[543] Ely DP. Trends in Educational Technology. ERIC Rep. ED346850. 1992. [Online]. Available from: https://files.eric.ed.gov/fulltext/ED346850.pdf

[544] Onele NO. Virtual reality: A tool for improving the teaching and learning of technology education. In: Virtual Reality, Ch. 5. London, UK: IntechOpen; 2020. pp. 1-15

[545] Rodríguez JL. Virtual reality in the classroom: A difficult but exciting

adventure for teachers and students. Frontiers in Education. 2024;**9**:1294715. DOI: 10.3389/feduc.2024.1294715

[546] Chimakurthi VNSS. Emerging of virtual reality (VR) technology in education and training. Asian Journal of Humanity, Art and Literature. 2018;**5**(2):157. DOI: 10.18034/ajhal.v5i2.606

[547] Castro MP, Garduño HS. Beyond traditional classrooms: Comparing virtual reality applications and their influence on students' motivation. Education in Science. 2024;**14**(9):963. DOI: 10.3390/educsci14090963

[548] Vert S, Andone D. Virtual reality authoring tools for educators. In: Proc. ITM Web Conf. Paris, France; 2019, Art. no. 3008

[549] Marougkas A et al. Virtual reality in education: A review of learning theories, approaches and methodologies for the last decade. Electronics. 2023;**12**(13):2832. DOI: 10.3390/electronics12132832

[550] Asad MM et al. Virtual reality as pedagogical tool to enhance experiential learning: A systematic literature review. Education Research International. 2021;**2021**:1. DOI: 10.1155/2021/7061623

[551] Schuster CM, Moloney MJ. The future of virtual reality in education. In: Proc. ACM Conf. Future Educ. Boston, MA, USA; 2021. pp. 85-89

[552] Kirubakaran SJ et al. A feasible approach to smart remote health monitoring: Subscription-based model. Frontiers in Public Health. 2023;**11**:1150455. DOI: 10.3389/fpubh.2023.1150455

[553] Man J et al. Innovative analysis of higher vocational education model based on virtual reality technology. Journal of Physics Conference Series. 2020;**1533**(2):22097. DOI: 10.1088/1742-6596/1533/2/022097

[554] Yang H. Explore how artificial intelligence and VR technology will change the development of future education. Journal of Physics Conference Series. 2021;**1744**(4):42146. DOI: 10.1088/1742-6596/1744/4/042146

[555] Kavanagh S et al. A systematic review of virtual reality in education. International Journal of Virtual and Augmented Reality. 2017;**10**(2):85. [Online]. Available from: https://files.eric.ed.gov/fulltext/EJ1165633.pdf

[556] Chen M-F et al. Design and evaluation of a remote synchronous gamified mathematics teaching activity that integrates multi-representational scaffolding and a mind tool for gamified learning. Education and Information Technologies. 2023;**28**(10):13207. DOI: 10.1007/s10639-023-11708-6

[557] Mendolia-Moore T. The role of virtual reality in distance learning: An examination of the efficacy on student learning outcomes, social presence, and collaborative learning [Ph.D. dissertation]. Univ. XYZ; 2023

[558] Wang Y. The influence of virtual reality technology on the cultivation of agricultural students. Mobile Information Systems. 2021;**2021**:1. DOI: 10.1155/2021/7699106

[559] Lee H et al. Virtual reality metaverse system supplementing remote education methods: Based on aircraft maintenance simulation. Applied Sciences. 2022;**12**(5):2667. DOI: 10.3390/app12052667

[560] Peruzzini M et al. Evaluating the impact of virtual reality on mechanical

design education. In: Proc. ASME/IEEE Int. Conf. Mechatron. Embed. Syst. Appl. San Diego, CA, USA; 2009. pp. 107-114

[561] Schott C, Marshall S. Virtual reality for experiential education: A user experience exploration. Australasian Journal of Educational Technology. 2020;**36**(4):96. DOI: 10.14742/ajet.5166

[562] Dobrova VV et al. Language education in the age of globalization and innovation. SHS Web of Conferences. 2019;**71**:5010. DOI: 10.1051/shsconf/20197105010

[563] Aydın S, Aktaş B. Developing an integrated VR infrastructure in architectural design education. Frontiers in Robotics and AI. 2020;7:495468. DOI: 10.3389/frobt.2020.495468

[564] Merchant E et al. Effectiveness of virtual reality-based instruction on students' learning outcomes in K-12 and higher education: A meta-analysis. Computers in Education. 2013;**70**:29. DOI: 10.1016/j.compedu.2013.07.033

[565] Analyti E et al. Virtual reality in education: Transforming learning through immersive technology. Technium Education and Humanities. 2024;**10**:1. DOI: 10.47577/teh.v10i.11766

[566] Neden J. Sustainable, agile technology navigation accessing virtuality for real-world learning: A SATNAV for social work educators. Social Work Education. 2020;**41**(2):195. DOI: 10.1080/02615479.2020.1821636

[567] Turienzo J. A transversal and practical education as a business success factor: Literature review of learning process of basic design through ICT tools. Journal of Management and Business Education. 2024;7(1):70. DOI: 10.35564/jmbe.2024.0005

[568] Hamza TS, Hassan DK. Consequential creativity. International Journal of Technology and Design Education. 2015;**26**(4):587. DOI: 10.1007/s10798-015-9321-4

[569] Makaklı ES. STEAM approach in architectural education. SHS Web of Conferences. 2019;**66**:1012. DOI: 10.1051/shsconf/20196601012

[570] Mo NN et al. Teaching mode reform of architectural design based on virtual reality. Applied Mechanics and Materials. 2014;**584-586**:284. DOI: 10.4028/www.scientific.net/AMM.584-586.284

[571] Karimi Z, Uddin MS. A missing link: Thinking | making | presenting. SHS Web of Conferences. 2019;**64**:2017. DOI: 10.1051/shsconf/20196402017

[572] Sanchez-Sepulveda MV et al. Visual technologies for urban design competences in architecture education. In: Proc. ACM Int. Conf. Urban des. Educ. New York, NY, USA; 2019. pp. 726-730

[573] Zalloom B. Increasing creativity and community responsibility through the interactive learning at the schools of architecture in Jordan. IOP Conference Series: Materials Science and Engineering. 2019;**471**:82065. DOI: 10.1088/1757-899X/471/8/082065

[574] Khadri HO. Becoming future-proof STEM teachers for enhancing sustainable development: A proposed general framework for capacity-building programs in future studies. Prospects. 2022;**52**:421. DOI: 10.1007/s11125-021-09588-0

[575] Zhanqiang M. Strategies for cultivating creative thinking ability of students majoring in educational technology. Frontiers in Educational Research. 2023;**6**(19). DOI: 10.25236/fer.2023.061911

[576] Ramaraj A, Nagammal J. Examining the plausibility of fostering creativity through puzzles in architectural education: An exploratory sequential study. Thinking Skills and Creativity. 2017;**24**:48. DOI: 10.1016/j.tsc.2017.02.001

[577] Majithia RK. What's next in design education? Transforming role of a designer and its implications in preparing youth for an ambiguous and volatile future. The Design Journal. 2017;**20**:35. DOI: 10.1080/14606925.2017.1352676

[578] Bosman L. From doing to thinking: Developing the entrepreneurial mindset through scaffold assignments and self-regulated learning reflection. Open Education Studies. 2019;**1**(1):106. DOI: 10.1515/edu-2019-0007

[579] Aguayo C et al. Ethical enactivism for smart and inclusive STEAM learning design. Heliyon. 2023;**9**(9):e19205. DOI: 10.1016/j.heliyon.2023.e19205

[580] Fazylzianova GI, Balalov VV. Training in the digitalization of education: Advanced learning technologies and design thinking methodology. In: Proc. New Silk Road Bus. Coop. Prospective Econ. Dev. Conf. Almaty, Kazakhstan; 2020. pp. 450-457

[581] Oppert ML et al. A mixed-methods study of creative problem solving and psychosocial safety climate: Preparing engineers for the future of work. Frontiers in Psychology. 2022;**12**:759226. DOI: 10.3389/fpsyg.2021.759226

[582] Suastrawan KE et al. The effectiveness of science e-modules for class VII junior high schools based on socioscientific issues to improve students' critical thinking skills. Journal of Science Education Research. 2021;**5**(2):1. DOI: 10.21587/jser.v5i2.42877

[583] Iringan EM. Instructional exposure of senior high school students to approaches that promote critical thinking and problem-solving skills. Journal of Asian Research. 2021;**5**(1):1. DOI: 10.22158/jar.v5n1p1

[584] Hanum L et al. Development of learning devices based on ethnoscience project based learning to improve students' critical thinking skills. Jurnal Pendidikan Sains Indonesia. 2023;**11**(2):288. DOI: 10.24815/jpsi.v11i2.28294

[585] Himmatussolihah H et al. Critical thinking skills of 10th grade students and the effect on learning achievement. In: Proc. 3rd Int. Conf. Learn. Innov. Qual. Educ. (ICLIQE). Malang, Indonesia; 2020. pp. 1-6

[586] Muntaha M et al. Content analysis of critical-rand creative-thinking skills in middle-school science books on environmental pollution material. Journal of Physics Conference Series. 2021;**1806**(1):12138. DOI: 10.1088/1742-6596/1806/1/012138

[587] Kreijns K et al. Adopting the integrative model of behaviour prediction to explain teachers' willingness to use ICT: A perspective for research on teachers' ICT usage in pedagogical practices. Technology, Pedagogy and Education. 2013;**22**(1):55. DOI: 10.1080/1475939X.2012.754371

[588] Widiatmo T et al. Analysis of validation of instruments to measure student's critical thinking ability and science literation. Jurnal Penelitian Pendidikan IPA. 2019;**5**(2):212. DOI: 10.29303/jppipa.v5i2.272

[589] Dumitru D, Halpern DF. Critical thinking: Creating job-proof skills for the future of work. Journal of Intelligence. 2023;**11**(10):194. DOI: 10.3390/jintelligence11100194

[590] Roviati E et al. Prospective biology teachers' critical thinking skills in microbiology argument-based inquiry laboratory activities. Journal of Physics Conference Series. 2019;**1157**:22108. DOI: 10.1088/1742-6596/1157/2/022108

[591] Jamil M. Secondary school science teachers' practices for the development of critical thinking skills: An observational study. Journal of Development and Social Sciences. 2021;**2**:259. DOI: 10.47205/jdss.2021(2-iv)22

[592] Aliftika O et al. Project based blended learning and independent learning on critical thinking skill. Journal of Physics Conference Series. 2021;**2019**(1):12051. DOI: 10.1088/1742-6596/2019/1/012051

[593] English LD. Ways of thinking in STEM-based problem solving. ZDM Mathematics Education. 2023;**55**(7):1219. DOI: 10.1007/s11858-023-01474-7

[594] Saputri AC et al. Critical thinking skills profile of senior high school students in biology learning. Journal of Physics Conference Series. 2018;**1006**:12002. DOI: 10.1088/1742-6596/1006/1/012002

[595] Chusni MM et al. The potential of discovery learning models to empower students' critical thinking skills. Journal of Physics Conference Series. 2020;**1464**(1):12036. DOI: 10.1088/1742-6596/1464/1/012036

[596] Zakiyah R et al. The influence of science, technology, engineering, mathematic (STEM) based biology learning through inquiry learning models towards students' critical thinking skills and mastery of biological concepts. AIP Conf. Proc. 2021;**2330**:30061. DOI: 10.1063/5.0043361

[597] Albab AU, Wangguway Y. Profile of students' creative and innovative thinking in solving open-ended mathematics problems about the coffee plantation. Journal of Physics Conference Series. 2020;**1538**(1):12071. DOI: 10.1088/1742-6596/1538/1/012071

[598] Miskiah M et al. The effects of blended learning on elementary school students' creativity and activeness. Universal Journal of Educational Research. 2020;**8**(9):3958. DOI: 10.13189/ujer.2020.080920

[599] Anggraeni P et al. Why 6 Cs? The urgency of learning at elementary school. In: Proc. Adv. Soc. Sci. Educ. Hum. Res. Conf. Jakarta, Indonesia; 2022. pp. 1-5

[600] Pellegrino JW, Hilton ML. Education for Life and Work: Developing Transferable Knowledge and Skills in the 21st Century. Washington, DC, USA: National Academies Press; 2012

[601] Gasaymeh A-MM, AlMohtadi RM. College of education students' perceptions of their computational thinking proficiency. Frontiers in Education. 2024;**9**:1478666. DOI: 10.3389/feduc.2024.1478666

[602] Nufus H et al. Analyzing the students' mathematical creative thinking ability in terms of self-regulated learning: How do we find what we are looking for? Heliyon. 2024;**10**(3):e24871. DOI: 10.1016/j.heliyon.2024.e24871

[603] Johnson P. The 21st century skills movement. Educational Leadership. 2009;**67**(1):11. [Online]. Available from: https://eric.ed.gov/?id=EJ855078

[604] Li Y et al. Enhancing early childhood education: Integrating theory and practice. Communications in Humanities Research. 2024;**32**(1):202. DOI: 10.54254/2753-7064/32/20240071

[605] Lee SJ, Kwon K. A systematic review of AI education in K-12 classrooms

from 2018 to 2023: Topics, strategies, and learning outcomes. Computers and Education: Artificial Intelligence. 2024;**6**:100211. DOI: 10.1016/j.caeai.2024.100211

[606] Faresta RA et al. Exploring the potential of virtual reality (VR) in developing students' thinking skills: A narrative review of the last five years. International Journal of Essential Competencies in Education. 2024;**3**(2):217. DOI: 10.36312/ijece.v3i2.2407

[607] Camperos JAG et al. SpaceRaceEdu: Developing an educational multi-player videogame for self-study and assessment. arXiv:2410.13875 [cs]. 2024

[608] Khawaja E, Ehsan H. Work in progress: A survey of artificial intelligence educational resources for pre-college education. In: Proc. ASEE Conf. Atlanta, GA, USA; 2024, Art. no. 1-2--44153

[609] Ferrell V, Tharpe AS. Enhancing rural science education through school district–university partnership. Education in Science. 2024;**14**(7):712. DOI: 10.3390/educsci14070712

[610] Alotaibi NS. The impact of AI and LMS integration on the future of higher education: Opportunities, challenges, and strategies for transformation. Sustainability. 2024;**16**(23):10357. DOI: 10.3390/su162310357

[611] Çelik F, Baturay MH. Technology and innovation in shaping the future of education. Smart Learning Environments. 2024;**11**(1):1. DOI: 10.1186/s40561-024-00339-0

[612] Pierce GL, Cleary PF. The persistent educational digital divide and its impact on societal inequality. PLoS One. 2024;**19**(4):e0286795. DOI: 10.1371/journal.pone.0286795

[613] Huang R et al. Global understanding of smart education in the context of digital transformation. Open Praxis. 2024;**16**(4):663. DOI: 10.55982/openpraxis.16.4.761

[614] Gao A, Chen W. The association between internet use and cognitive ability among rural left-behind children in China. Frontiers in Public Health. 2024;**11**:1341298. DOI: 10.3389/fpubh.2023.1341298

[615] Gilliard E et al. Protecting Africa's future: Cybersecurity strategies for child safety, learning, and skill acquisition in Tanzania. arXiv:2409.13159 [cs]. 2024

[616] Marín JAM et al. Computational thinking and programming with Arduino in education: A systematic review for secondary education. Heliyon. 2024;**10**(8):e29177. DOI: 10.1016/j.heliyon.2024.e29177

[617] Bachtsis R et al. Challenges faced by students with special needs in primary education during online teaching. Education in Science. 2024;**14**(3):220. DOI: 10.3390/educsci14030220

[618] Li Y et al. The impact of digital educational games on student's motivation for learning: The mediating effect of learning engagement and the moderating effect of the digital environment. PLoS One. 2024;**19**(1):e0294330. DOI: 10.1371/journal.pone.0294330

[619] Parveen DS, Ramzan SI. The role of digital technologies in education: Benefits and challenges. International Research Journal on Advanced Engineering and Management. 2024;**2**(6):2029. DOI: 10.47392/irjaem.2024.0299

[620] Kim YE et al. Enabling creative collaboration for all levels of learning. Proceedings of the National Academy of

Sciences. 2019;**116**(6):1878. DOI: 10.1073/pnas.1808678115

[621] Alphrazy R, Octavia B. Effectiveness of student teams achievement division and scramble combined model on collaborative skills and conceptual knowledge mastery of class X SMAN 1 semparuk on bacteria. Jurnal Penelitian Pendidikan IPA. 2023;**9**(6):4783. DOI: 10.29303/jppipa.v9i6.3544

[622] Jardim J. Entrepreneurial skills to be successful in the global and digital world: Proposal for a frame of reference for entrepreneurial education. Education in Science. 2021;**11**(7):356. DOI: 10.3390/educsci11070356

[623] Higgins S. Critical thinking for 21st-century education: A cyber-tooth curriculum? Prospects. 2014;**44**(4):559. DOI: 10.1007/s11125-014-9323-0

[624] Kain C et al. Mapping the landscape: A scoping review of 21st century skills literature in secondary education. Teaching and Teacher Education. 2024;**151**:104739. DOI: 10.1016/j.tate.2024.104739

[625] Rizaldi DR et al. The correlation of digital literacy and STEM integration to improve Indonesian students' skills in 21st century. International Journal of Asian Education. 2020;**1**(2):73. DOI: 10.46966/ijae.v1i2.36

[626] Krajcik J, Delen İ. Engaging learners in STEM education. Estonian Journal of Education. 2017;**5**(1):35. DOI: 10.12697/eha.2017.5.1.02b

[627] Latip A, Hardinata A. Implementation of STEM-robotics as high school intra-curricular. Thabiea: Journal of Natural Science Teaching. 2020;**3**(1):11. DOI: 10.21043/thabiea.v3i1.6770